Bioremediation of Inorganics

BIOREMEDIATION

The *Bioremediation* series contains collections of articles derived from many of the presentations made at the First, Second, and Third International In Situ and On-Site Bioreclamation Symposia, which were held in 1991, 1993, and 1995 in San Diego, California.

First International In Situ and On-Site Bioreclamation Symposium

- **1(1)** *On-Site Bioreclamation: Processes for Xenobiotic and Hydrocarbon Treatment*
- **1(2)** *In Situ Bioreclamation: Applications and Investigations for Hydrocarbon and Contaminated Site Remediation*

Second International In Situ and On-Site Bioreclamation Symposium

- **2(1)** *Bioremediation of Chlorinated and Polycyclic Aromatic Hydrocarbon Compounds*
- **2(2)** *Hydrocarbon Bioremediation*
- **2(3)** *Applied Biotechnology for Site Remediation*
- **2(4)** *Emerging Technology for Bioremediation of Metals*
- **2(5)** *Air Sparging for Site Bioremediation*

Third International In Situ and On-Site Bioreclamation Symposium

- **3(1)** *Intrinsic Bioremediation*
- **3(2)** *In Situ Aeration: Air Sparging, Bioventing, and Related Remediation Processes*
- **3(3)** *Bioaugmentation for Site Remediation*
- **3(4)** *Bioremediation of Chlorinated Solvents*
- **3(5)** *Monitoring and Verification of Bioremediation*
- **3(6)** *Applied Bioremediation of Petroleum Hydrocarbons*
- **3(7)** *Bioremediation of Recalcitrant Organics*
- **3(8)** *Microbial Processes for Bioremediation*
- **3(9)** *Biological Unit Processes for Hazardous Waste Treatment*
- **3(10)** *Bioremediation of Inorganics*

Bioremediation Series Cumulative Indices: 1991-1995

For information about ordering books in the Bioremediation series, contact Battelle Press. Telephone: 800-451-3543 or 614-424-6393. Fax: 614-424-3819. Internet: sheldric@battelle.org.

Bioremediation of Inorganics

Edited by

Robert E. Hinchee and Jeffrey L. Means
Battelle Memorial Institute

David R. Burris
U.S. Air Force Armstrong Laboratory

Library of Congress Cataloging-in-Publication Data

Hinchee, Robert E.
Bioremediation of inorganics / edited by Robert E. Hinchee, Jeffrey L. Means, David R. Burris.
p. cm.
Includes bibliographical references and index.
ISBN 1-57477-011-X (hc : acid-free paper)
1. Metal wastes—Biodegradation—Congresses. 2. Inorganic compounds—Biodegradation—Congresses. 3. Bioremediation—Congresses. I. Hinchee, Robert E. II. Means, Jeffrey L. III. Burris, David R.
TD196.M4B57 1995
628.5′2—dc20 95-32251
CIP

Printed in the United States of America

Additional copies may be ordered through:
Battelle Press
505 King Avenue
Columbus, Ohio 43201, USA
1-614-424-6393 or 1-800-451-3543
Fax: 1-614-424-3819
Internet: sheldric@battelle.org

CONTENTS

FOREWORD

This book and its companion volumes (see overleaf) comprise a collection of papers derived from the Third International In Situ and On-Site Bioreclamation Symposium, held in San Diego, California, in April 1995. The 375 papers that appear in these volumes are those that were accepted after peer review. The editors believe that this collection is the most comprehensive and up-to-date work available in the field of bioremediation.

Significant advances have been made in bioremediation since the First and Second Symposia were held in 1991 and 1993. Bioremediation as a whole remains a rapidly advancing field, and new technologies continue to emerge. As the industry matures, the emphasis for some technologies shifts to application and refinement of proven methods, whereas the emphasis for emerging technologies moves from the laboratory to the field. For example, many technologies that can be applied to sites contaminated with petroleum hydrocarbons are now commercially available and have been applied to thousands of sites. In contrast, there are as yet no commercial technologies commonly used to remediate most recalcitrant compounds. The articles in these volumes report on field and laboratory research conducted both to develop promising new technologies and to improve existing technologies for remediation of a wide spectrum of compounds.

The editors would like to recognize the substantial contribution of the peer reviewers who read and provided written comments to the authors of the draft articles that were considered for this volume. Thoughtful, insightful review is crucial for the production of a high-quality technical publication. The peer reviewers for this volume were:

Aurel Acher, *Volcany Center* (Israel)
Michael D. Aitken, *University of North Carolina*
Erik Arvin, *Technical University of Denmark*
David Bass, *Groundwater Technology Inc.*
Jason Caplan, *En Solve Biosystems, Inc.*
Kung-Hui Chu, *University of California, Berkeley*
Bob Fallon, *DuPont Co.*
Brian R. Folsom, *Envirogen, Inc.*
Eric A. Foote, *Battelle Columbus*
William Frankenberger, *University of California*
Michael Gunderson, *HAZWRAP*
Jennie Hunter-Cevera, *Lawrence Berkeley Laboratory*
Yacov Kanfi, *Israel Ministry of Agriculture*
Michiel J.J. Kotterman, *Agricultural University of Wageningen* (The Netherlands)
Laurie Lapat-Polasko, *Woodward-Clyde Consultants*
Kun Mo Lee, *Ajou University* (Korea)
Tony Lieberman, *ESE Biosciences, Inc.*

Michael McFarland, *Utah State University*
Nigel Quinn, *Lawrence Berkeley Laboratory*
JoAnn Radway, *Westinghouse Savannah River Co.*
Alfons J.M. Stams, *Agricultural University of Wageningen* (The Netherlands)
Keith E. Stormo, *Innovative BioSystems, Inc.*
Gerald Strandberg, *Oak Ridge National Laboratory*
Janet Strong-Gunderson, *Oak Ridge National Laboratory*
Per Sveum, *SINTEF Applied Chemistry* (Norway)
C. Michael Swindoll, *DuPont Co.*
Edward Wilde, *Westinghouse Savannah River Co.*
Thomas Wildeman, *Colorado School of Mines*
Kenneth J. Williamson, *Oregon State University*

Finally, I want to recognize the key members of the production staff, who put forth significant effort in assembling this book and its companion volumes. Carol Young, the Symposium Administrator, was responsible for the administrative effort necessary to produce the ten volumes. She was assisted by Gina Melaragno, who tracked draft manuscripts through the review process and generated much of the correspondence with the authors, co-editors, and peer reviewers. Lynn Copley-Graves oversaw text editing and directed the layout of the book, compilation of the keyword indices, and production of the camera-ready copy. She was assisted by technical editors Bea Weaver and Ann Elliot. Loretta Bahn was responsible for text processing and worked many long hours incorporating editors' revisions, laying out the camera-ready pages and figures, and maintaining the keyword list. She was assisted by Sherry Galford and Cleta Richey; additional support was provided by Susan Vianna and her staff at Fishergate, Inc. Darlene Whyte and Mike Steve proofread the final copy. Judy Ward, Gina Melaragno, Bonnie Snodgrass, and Carol Young carried out final production tasks. Karl Nehring, who served as Symposium Administrator in 1991 and 1993, provided valuable insight and advice.

The symposium was sponsored by Battelle Memorial Institute with support from many organizations. The following organizations cosponsored or otherwise supported the Third Symposium.

Ajou University–College of Engineering (Korea)
American Petroleum Institute
Asian Institute of Technology (Thailand)
Biotreatment News
Castalia
ENEA (Italy)
Environment Canada
Environmental Protection
Gas Research Institute
Groundwater Technology, Inc.

Institut Français du Pétrole
Mitsubishi Corporation
OHM Remediation Services Corporation
Parsons Engineering Science, Inc.
RIVM–National Institute of Public Health and the Environment (The Netherlands)
The Japan Research Institute, Limited
Umweltbundesamt (Germany)
U.S. Air Force Armstrong Laboratory–Environics Directorate
U.S. Air Force Center for Environmental Excellence
U.S. Department of Energy Office of Technology Development (OTD)
U.S. Environmental Protection Agency
U.S. Naval Facilities Engineering Services Center
Western Region Hazardous Substance Research Center–Stanford and Oregon State Universities

Neither Battelle nor the cosponsoring or supporting organizations reviewed this book, and their support for the Symposium should not be construed as an endorsement of the book's content. I conducted the final review and selection of all papers published in this volume, making use of the essential input provided by the peer reviewers and other editors. I take responsibility for any errors or omissions in the final publication.

Rob Hinchee
June 1995

An Overview of the Bioremediation of Inorganic Contaminants

Harvey Bolton, Jr., and Yuri A. Gorby

ABSTRACT

Bioremediation, or the biological treatment of wastes, usually is associated with the remediation of organic contaminants. Similarly, there is an increasing body of literature and expertise in applying biological systems to assist in the bioremediation of soils, sediments, and water contaminated with inorganic compounds including metals, radionuclides, nitrates, and cyanides. Inorganic compounds can be toxic both to humans and to organisms used to remediate these contaminants. However, in contrast to organic contaminants, most inorganic contaminants cannot be degraded, but must be remediated by altering their transport properties. Immobilization, mobilization, or transformation of inorganic contaminants via bioaccumulation, biosorption, oxidation, reduction, methylation, demethylation, metal-organic complexation, ligand degradation, and phytoremediation are the various processes applied in the bioremediation of inorganic compounds. This paper briefly describes these processes, referring to other contributors in this book as examples when possible, and summarize the factors that must be considered when choosing bioremediation as a cleanup technology for inorganics. Understanding the current state of knowledge as well as the limitations for bioremediation of inorganic compounds will assist in identifying and implementing successful remediation strategies at sites containing inorganic contaminants.

INTRODUCTION

Metals are essential to life and are used as micronutrients by humans, plants, and microorganisms. However, when the concentration of micronutrients in the environment becomes higher than is required by living organisms, they can become toxic and are then considered to be contaminants. The consequences of soils, sediments, or waters that have become contaminated with inorganic compounds are the hazards they pose to human health and ecological processes. Biological processes can play a role in the treatment or remediation of soils, sediments, and waters containing inorganic contaminants such as metals, radionuclides, nitrates, and cyanides. This is because select biotic systems (e.g., microorganisms and plants) have the ability to render inorganic contaminants present at high concentrations in their environment nontoxic, presumably as a defense mechanism. These organisms and the processes they catalyze can be used to remediate sites contaminated with inorganic compounds.

The mechanisms used to reduce the toxicity of inorganic contaminants include immobilization, mobilization, and transformation. Examples of biotically mediated processes that utilize these mechanisms include bioaccumulation, biosorption, oxidation, reduction, methylation, demethylation, metal-organic complexation, ligand degradation, and phytoremediation. Understanding how inorganic contaminants interact with biological systems and how biological systems adapt and respond to inorganic contaminants and their environment is necessary for understanding biological treatment options. Adaptation to the environment and the presence of inorganic contaminants (e.g., changing contaminant levels and extremes of pH, temperature, salinity, and ionic strength) can be an advantage in using biotic systems for the treatment of inorganic contaminants (Whitlock 1990). Initially changes such as swings in concentration of inorganic or organic constituents or changes in pH or Eh may adversely affect the process, but resistance to these changes usually improves over time.

This paper briefly describes these various processes and discusses how they may be applied as remediation strategies using papers from this book as examples. It is not intended to be a comprehensive review of the topic. Each of the following sections described herein represents a broad topic to which entire chapters and even books have been devoted (Beveridge 1989b; Ehrlich and Brierley 1990; Volesky 1990a). Rather, the purpose of this paper is to introduce the reader to the bioremediation of inorganic contaminants and provide a background for later sections of this book. Understanding the current state of knowledge as well as the limitations to bioremediation of inorganic compounds will assist in developing and implementing effective remediation strategies at sites contaminated with inorganic compounds.

BIOACCUMULATION AND BIOSORPTION OF INORGANIC CONTAMINANTS

Microorganisms assimilate and concentrate cationic nutrients from solution, and they may use these same mechanisms to concentrate metals and radionuclides. An "active" process called bioaccumulation takes place when metabolic energy is used for the assimilation of the inorganic compound. Uptake that does not require metabolic energy is termed biosorption and usually involves the complexation of inorganic contaminants by ligands or functional groups on the outside surface of the cell. Microorganisms sequester metals internally by complexing them with various cytoplasm ligands including polyphosphates or proteins.

Bioaccumulation and biosorption of a wide range of cationic metals and radionuclides has been demonstrated for bacteria, fungi, and algae (Poole and Gadd 1989; Volesky 1990a). Microorganisms often exhibit some selectivity in the binding of inorganic contaminants, more so than most synthetic chemical sorbents. The mechanisms of metal biosorption and bioaccumulation by microorganisms is usually dictated by the preference of metals for different ligand binding sites. As an example, Cd can be complexed to polyphosphates (Higham et al. 1985, 1986b) and to sulfur-containing proteins in the cytoplasm (Higham et al. 1984; Higham et al. 1986a; Khazaeli and Mitra 1981; Mitra et al. 1975). The metal can also precipitate outside the cell as Cd phosphate (Aiking et al. 1984; Macaskie and Dean 1984; Macaskie et al. 1987) or as Cd sulfide (Aiking et al. 1982). These inorganic precipitates form because the P and S ligands are preferred by Cd^{2+}, which is a soft acid. In some instances, the amount of metal biosorbed by the exterior of the cell exceeds that which would be can be predicted from the charge density of the cell wall. It has been hypothesized that a two-stage process exists in which the metal interacts with the charged moieties on the cell wall followed by the deposition of more metal using the initially bound metal as a nucleation site (Beveridge 1989a).

In addition to cell walls, other extracellular material can promote biosorption of metals. Microorganisms produce polymers on the outside of the cell which allow them to attach to surfaces. These exopolymers, also called capsules or slime layers, can sorb radionuclides and metals under natural conditions (Ferris 1989; Mittelman and Geesey 1985) as well as in waste treatment facilities (Gadd and White 1993). Exopolymers are acidic in nature with uronic acids and other acidic functional groups consisting of substituted sugars providing carboxylate residues and oxygen atoms for radionuclide and/or metal complexation and coordinate binding (Geesey and Jang 1989). The production of exopolymers can be controlled by the cultural conditions including the C and N source and the presence, absence, or concentration of nutrient and contaminant ions.

Metals can be precipitated at the cell surface as metal-phosphates or metal-sulfides to immobilize metals and radionuclides. As an example, Cd phosphate precipitation was promoted by a *Citrobacter* sp. (Macaskie and Dean 1984; Macaskie et al. 1987). A phosphatase bound to the cell wall of *Citrobacter* sp. produces HPO_4^{2-} from glycerol 2-phosphate, which then precipitates the Cd as cell-bound Cd phosphate (Macaskie and Dean 1984; Macaskie et al. 1987). This inorganic phosphate-generating system also actively precipitates uranium as uranyl phosphate (Macaskie et al. 1992) and has been used for the treatment of metal-containing wastestreams (Macaskie 1990). *Pseudomonas fluorescens* and a *Citrobacter* sp. have also been shown to accumulate Pb as Pb phosphates on the cell surface (Aickin and Dean 1979). Cd and Cu sulfides precipitated at the cell wall of a *Klebsiella aerogenes* (Aiking et al. 1982; Aiking et al. 1984) and a *Mycobacterium scrofulaceum* (Erardi et al. 1987). A more detailed discussion of the precipitation of metals by sulfide generation is presented in the Oxidation and Reduction section of this paper.

Two papers in this book address biosorption processes that assist in the bioremediation of inorganic contaminants. The goal of Wang et al. (1995) was to identify plant and algal species that might be effective at removing inorganic contaminants from wastestreams. They tested six vascular plants, two macroalgae, and five microalgae for Cd and Zn (1 mg/L) accumulation. These organisms offer potential for the removal of metals from solution via bioaccumulation and biosorption mechanisms. Lu and Wilkins (1995) investigated the metal biosorptive properties of *Saccharomyces cerevisiae* (yeast) immobilized in an alginate matrix. *S. cerevisiae* is available in large quantities from the food industry. Immobilization of the organism in an alginate matrix may enhance its application by allowing biomass to be used as an ion exchange resin. No decrease in the biosorptive capacity was found for the immobilized yeast over seven acid treatment cycles to remove the sorbed metals and restore charged sites, demonstrating that the yeast-alginate bead system can be used repeatedly.

Immobilization of biomass, as done by Lu and Wilkins (1995), may enhance the utility of biosorption by adding mechanical strength and uniformity in size, and by allowing biosorbents to be used as ion exchange resins in existing treatment scenarios. Support matrices that can be used for biomass immobilization include agar, alginates, cellulase, polyacrylamide, cross-linked ethyl acrylate-ethylene glycol dimethylacrylate, and the cross-linking reagents toluene diisocyanate and glutaraldehyde (Gadd and White 1993).

Commercially available biosorbents have been discussed previously (Brierley 1990; Volesky 1990b). These processes use microorganisms treated, via proprietary procedures, to enhance their removal of metals from aqueous systems. Examples include the BIOCLAIM process using *Bacillus* sp., the AlgaSORB™ process using algae, the BIO-FIX process using sphagnum peat moss, algae, yeast, bacteria, and/or aquatic microbiota (Brierley 1990), and B. V. SORBEX, Inc., which uses different types of microbial biomass (Volesky 1990b). These systems can be used in batch, continuous-flow, fixed-bed, fluidized-bed, and various multiple-bed reactors (Volesky 1990b). A main advantage of using biologically derived biomass for biosorption is the low costs of production (i.e., growing the organisms). Also, microorganisms may be altered either through cultural conditions or through genetic manipulation to overproduce selected ligands necessary to remediate inorganic contaminants.

Bioaccumulation/biosorption to assist in the remediation of inorganic contaminants has been used at the Homestake Mine in South Dakota. At this facility, rotating biological contactors containing *Pseudomonas* sp. were used to treat the wastewater for the removal of Cu and Fe, the primary metal contaminants in the influent wastewater. The removal of Cu and Fe reached 95 to 98% of the influent concentration. This facility can treat 4 million gal (15 million L) of wastewater a day (Whitlock 1990). One precaution with the facility is that the biomass can become saturated with the target metals and start desorbing these metals so that the effluent has a higher concentration than the influent. This is most likely because of displacement of the Cu and Fe from the saturated biomass by other cations in the wastewater. Thus, the biomass must be maintained at a certain age for the facility to remove Cu and Fe efficiently (Whitlock 1990).

OXIDATION AND REDUCTION

Microorganisms can influence inorganic contaminant solubility and mobility by direct or indirect oxidation (removing electrons and increasing metal valence) or reduction (adding electrons and decreasing metal valence). The solubility and toxicity of metals can vary depending upon their valence. As an example, Cr(III) is relatively nontoxic and nonmobile in the environment. The oxidized form Cr(VI) (chromate, CrO_4^{2-}) by comparison is mobile and toxic. Direct enzymatic oxidation or reduction reactions significantly alter the solubility of a wide range of ions, such as Cr, Fe, Mn, Hg, NO_3-N, Se, and U (Lovley 1994). Indirect processes that remove contaminant metals from solution include precipitation of the metal sulfides. The application of microbially catalyzed oxidation/reduction processes has real potential for remediating surface waters, groundwaters, and industrial wastestreams contaminated with inorganic compounds. There are multiple examples in this book that describe microbially catalyzed oxidation/ reduction processes as a method for bioremediating inorganic contaminants. Here we describe a range of direct and indirect reduction/oxidation processes that are responsible for precipitating, volatilizing, and solubilizing inorganic contaminants.

Direct Processes: Enzymatic Reduction

A diverse group of bacteria can couple the oxidation of reduced organic matter or hydrogen to the reduction of Fe(III) or Mn(IV) (Balashova and Zavarzin 1980; Lovley 1994; Obuekwe and Westlake 1982). In anaerobic, nonsulfate-containing environments, dissimilatory metal-reducing bacteria are responsible for the reduction and mobilization of Fe(II) and Mn(II) from oxide minerals, the oxidation of nonfermentable organic matter, and the degradation of a variety of aromatic contaminants. Detailed descriptions of the ecological and physiological significance of dissimilatory metal-reducing bacteria are provided in a number of reviews (e.g., Lovley 1994).

In addition to Fe(III) and Mn(IV), metal-reducing bacteria are able to reduce a range of other multivalent metals. U(VI) to U(IV), Cr(VI) to Cr(III), Se(VI) to Se(0), and As(III) to As(0) are examples of metals that are amenable to reduction by bacteria and are soluble in their oxidized form but are insoluble when reduced. These metals form insoluble precipitates when reduced and thus may be removed from aqueous wastestreams or contaminated surface or groundwater.

Metal-reducing bacteria also can promote the solubilization of $PuO_{2(s)}$ in the presence of nitrilotriacetate (NTA, a synthetic chelator) (Rusin et al. 1994). The proposed mechanism is the reduction of insoluble Pu(IV) to soluble Pu(III), which is then complexed by the NTA. The NTA then oxidizes the Pu(III) to Pu(IV), resulting in the Pu(IV)-NTA complex, which is soluble. This process has potential for the solubilization and removal of Pu from contaminated soils and sediments.

Mercury can be removed from contaminated environments by aerobic bacteria that reduce soluble Hg(II) to volatile Hg(0). Reviews that describe the microbiology, biochemistry, and genetics of this process are available (Robinson and

Touvinen 1984; Silver and Walderhaug 1992; Summers and Barkay 1989). Exploitation of Hg(II)-reducing bacteria for treating contaminated environments has been considered by enhancing Hg reduction and volatilization (Barkay et al. 1991).

Three papers in this book investigated microbial selenium reduction as a bioremediation strategy. Selenium currently is a threat to agriculture and wildlife. The mobile and toxic selenate can be reduced to selenite and then to elemental Se, which is nonmobile and nontoxic and forms a red-colored precipitate. Selenite can support microbial growth as a terminal electron acceptor (Macy et al. 1989) or can be reduced by reactions independent of growth (Lortie et al. 1992). Owens et al. (1995) found 90% of 500 μg selenate/L was reduced by native sludge microorganisms in a conical-bottom upflow anaerobic sludge blanket reactor. The elemental Se precipitates and settles into the reactor sludge or may elute with the effluent and be concentrated in a secondary step. McCarty et al. (1995) investigated the reduction and methylation of selenium oxyanions by bacteria and detected volatile dimethyl selenide and dimethyl diselenide. Studies such as this assist in determining the mechanisms of metal reduction and how to optimize this process for its use in bioremediation. Finally, Garbisu et al. (1995) found two aerobic bacteria reduced selenite to elemental selenium. These organisms have the added advantage that there is no inhibition of selenite reduction by nitrate and sulfate.

Select microorganisms can utilize NO_3^- as a terminal electron acceptor to support growth and in the process convert NO_3^- to N_2. This microbially catalyzed process is biodenitrification. Two papers in this book discuss the use of biodenitrification to remediate environments contaminated with nitrate. Graydon et al. (1995) retrofitted septic tanks with recirculating trickling filters to enhance two microbial processes, first nitrification (NH_4^+ to NO_3^-) and then denitrification (NO_3^- to N_2). This system enhanced N removal from the septic systems by 66%. Schmidt and Ballew (1995) conducted bench-, pilot-, and field-scale studies using biodenitrification to remediate nitrate-containing water resulting from the processing of uranium and thorium ore. The results of the bench- and pilot-scale studies were used to develop the design parameters necessary for field implementation. This approach has been used to effectively treat 5.2 million gal (19.7 million L) of water contaminated with nitrate.

Indirect Reduction Processes

Metal contaminants can be removed from aqueous streams or waters by a number of indirect processes involving bacteria. Sulfide, formed by the enzymatic reduction of sulfate, can react with contaminants such as Ag, Cd, Ce, Ga, Hg, Ni, Pb, Se, Sr, and Zn to form insoluble metal sulfides. These sulfide precipitates are easily removed from the bulk aqueous phase by sedimentation, filtration, or centrifugation. Two papers in this book describe the application of bacterial sulfate reduction to remediate sulfate- and metal-contaminated water from mine drainage. Farmer et al. (1995) used sulfate-reducing bacteria to remove Ag, As, Cd, Mn, Ni, and Zn from mine drainage in continuous upflow bioreactors. The efficiency of Zn removal was 99% with an average influent concentration of 54 mg/L. Wildeman et al. (1995) conducted a pilot-scale passive treatment using sulfate reduction and sulfide precipitation of metals for treatment of seepage from mine tailings. Removal of Cu, Ni, and Zn from solution was maintained as long as organic matter, as a source of energy, and temperature were conducive to sulfate reduction.

Oxidative Processes

Select microorganisms called chemoautotrophs or chemolithotrophs gain their energy from the oxidation of reduced inorganic compounds including S and Fe. As an example, *Thiobacillus ferrooxidans* gains energy from the oxidation of mineral sulfides and Fe(II). The oxidation of sulfides produces sulfuric acid that solubilizes the ions of metals (e.g., Ag, Cd, Ce, Cu, Ga, Hg, Ni, Pb, Se, Sr, U, and Zn). This process is termed bioleaching (Tuovinen 1990) and is used to recover metals from sulfide ores. Bioleaching as a metal and mineral recovery process has been reviewed elsewhere (Ehrlich and Brierley 1990).

METHYLATION AND DEMETHYLATION

A number of microorganisms can methylate or demethylate metals. Methylation usually results in the volatilization of metals as described previously for the methylation of Se to volatile dimethyl selenide and dimethyl diselenide (McCarty et al. 1995). The methylation of metals may be a detoxification process to protect the organisms from the more toxic metal ion (Gadd 1993). The use of methylation to remediate the contamination of Se has potential because dimethyl selenide is much less toxic than inorganic Se species (Wilber 1980). However, the methylation of other metals can greatly enhance their toxicity to humans. As an example, methylmercury resulted in Minamata disease in Japan causing birth defects and was traced to the bioconcentration of CH_3Hg in fish that were consumed by the residents. Metals that can be methylated by microorganisms besides Se and Hg include As, Au, Pb, Pd, Pt, Sn, and Tl as described in the recent review by Gadd (1993).

One paper in this book investigated the bioleaching of As-contaminated soil by a mixed bacterial culture and volatilization of As by methanogenic bacteria (*Methanobacterium thermoautotrophicum*) (Bachofen et al. 1995). Anaerobic leaching resulted in mobilization of 45% of the As_2S_3 initially present at 8.3 g As/kg soil. *M. thermoautotrophicum* produced 25% volatile As species from the mobilized As. The volatile As species was not identified, but it is known that methanogens can methylate arsenate (McBride and Wolfe 1971). The use of an anaerobic bacterial metabolism in soil contaminated with arsenic with subsequent microbial volatilization of arsenic has potential as a remediation strategy either separately or combined with other applications. An important engineering aspect of the process is that volatile As species evolved from the contaminated material would have to be trapped in the off-gas.

Demethylation results in the breaking of the metal-methyl bond and usually results in charged metal. For example, CH_3Hg is degraded by organomercurial lyase to form Hg^{2+} and CH_4. The Hg^{2+} can then be reduced to Hg^0 which is volatile (Gadd 1993). One approach to remediate CH_3Hg-contaminated sites would be to enhance demethylation and Hg reduction and then to trap the volatile species. Saouter et al. (1994) proposed this strategy to remediate a Hg-contaminated pond near Oak Ridge, Tennessee. They proposed to add and stimulate native CH_3Hg-degrading and Hg-reducing bacteria to enhance $CH_3Hg \rightarrow Hg^{2+} \rightarrow Hg^0$. Shake flask and microcosms studies showed this approach has promise as long as the microbial degradation of CH_3Hg and the reduction of Hg^{2+} are enhanced over their inputs into the ecosystem or contaminated site under consideration.

METAL-ORGANIC COMPLEXATION

Microorganisms can produce extracellular metal-complexing agents or ligands that can enhance the solubility and mobility of metals in the environment. The complexation of the toxic radionuclides and/or metals may be a detoxification mechanism by reducing the aqueous concentration of the free inorganic contaminant. Also, complexing agents are produced by a wide variety of microorganisms to chelate iron and make it more available for growth (Lewin 1984). These chelating agents, called siderophores, may fortuitously chelate radionuclides and metals. Enhanced dissolution of PuO_2 occurred in the presence of Desferol, a modified siderophore (Barnhart et al. 1980). Siderophores from an *Anabaena* sp. protected the cell from Cu toxicity by complexing the Cu in solution and making it unavailable for cellular uptake (Clarke et al. 1987).

Organic compounds produced by microorganisms can also complex radionuclides and metals other than Fe. Microorganisms grown in the presence of Pu-DTPA (diethylenetriaminepentaacetic acid) produced soluble, negatively charged organic complexes of Pu with molecular weights greater than DTPA (Wildung and Garland 1980). Because the Pu initially was added as the DTPA complex to keep the Pu soluble, the microbially produced complex was able to out-compete the DTPA for the complexed Pu. *Pseudomonas aeruginosa* produced several previously uncharacterized chelating agents in the presence of high concentrations of soluble Th and U (Premuzic et al. 1986) and a *Cylindrocarpon* sp. produced an anionic Ni complex (Wildung et al. 1979). A marine bacterium *Vibrio alginolyticus* was found to produce extracellular proteins, which complexed and detoxified Cu (Harwood-Sears and Gordon 1990; Schreiber et al. 1990). These examples demonstrate that organisms can produce metal and radionuclide complexants, some of which may be useful for remediation applications. This is one area where future research is needed to determine what types, under what conditions, and how to apply microbially produced aqueous complexes to bioremediate soils and sediments contaminated with inorganic compounds via enhanced solubilization.

LIGAND DEGRADATION

Extractants are used commonly in waste processing, mining, and large-scale separations such as those employed by U.S. Department of Energy for the production of weapons-grade nuclear material. These include the synthetic chelating agents such as EDTA (ethylenediaminetetraacetic acid) and cyanide. The metal complexes that are formed result in enhanced metal mobility, and thus the metal is readily removed from the soil, sediment, ore, or other matrix. In contrast, degradation of the ligand could be used to decrease metal mobility and, in the case of cyanide, would also reduce the concentration of this toxic material.

EDTA is a chelating agent that is used in a variety of industrial applications and can enhance the solubility and mobility of metals in the environment. EDTA can be degraded by a bacterial consortium (Belly et al. 1975) and by several bacterial isolates (Lauff et al. 1990; Nörtmann 1992). The degradation of EDTA by strain BNC1 (Nörtmann 1992) depends on the chelated metal. In cases with Co(II), Co(III), or Ni(II), EDTA is not degraded (Bolton et al. 1995). Other metal-EDTA complexes such as Cu and Zn were degraded. These results suggest a possible bioremediation strategy would be to solubilize metals from contaminated sources using EDTA followed by degrading the metal-EDTA complex in a bioreactor.

Cyanides are present in various industrial wastestreams (e.g., metal plating and finishing) and from mining operations. Microorganisms can degrade free cyanide and some metal-cyanide complexes (Finnegan et al. 1991; Silva-Avalos et al. 1990). Although the mechanism of metal-cyanide complex degradation is unknown, in this book Aronstein et al. (1995) suggests that the organisms may be degrading the free cyanide dissociated from the metal-cyanide complex.

Several papers in this book describe the aerobic (Aronstein et al. 1995) and anaerobic (Nagle et al. 1995) degradation of cyanide as well as the anaerobic degradation of acetonitrile (Nagle et al. 1995) as a method of destroying these compounds. Aronstein et al. (1995) isolated aerobic microorganisms that could use cyanide as a sole N source from soil near a manufactured gas plant. Experiments were conducted at alkaline pH to prevent gaseous hydrogen cyanide generation during this process. Their results show that bacteria could degrade free cyanide, but they were unable to degrade the Fe-cyanide complex. Nagle et al. (1995) obtained mixed cultures of anaerobic bacteria from digested sewage sludge that degraded 85% of the cyanide and 83% of the acetonitrile, both present at starting concentrations of 50 mg/L. Using these organisms in a low-solids continuously stirred tank reactor, 71% of the acetonitrile was degraded at an input of 10 to 100 mg/L/day.

Degradation of cyanide has been scaled up to treat 4 million gal (15 million L) of wastewater per day from the Homestake Mine in South Dakota (Whitlock 1990). Rotating biological contactors containing *Pseudomonas* sp. were used to treat the wastewater containing 4 mg cyanide/L, reducing the effluent concentration to 0.06 mg/L.

PHYTOREMEDIATION

Phytoremediation is the use of plants to remediate contaminated soils or waters. The term is used here specifically for the remediation of inorganic contaminants but can also apply to the remediation of organic contaminants (Anderson and Coats 1994), primarily in the rhizosphere (Bolton et al. 1992). Phytoremediation of inorganic contaminants can be further divided into phytostabilization and phytoextraction (Cunningham et al. 1995). Phytostabilization is the use of plants to stabilize contaminated soil by decreasing wind and water erosion and also decreasing water infiltration and the subsequent leaching of contaminants. Phytoextraction is the removal of inorganic contaminants from aboveground portions of the plant. When the shoots and leaves are harvested, the inorganic contaminants are reclaimed or concentrated from the plant biomass.

Phytoremediation is an emerging technology that has recently received a lot of interest as demonstrated by three papers in this book (Cornish et al. 1995; Cunningham et al. 1995; Kelley and Guerin 1995). The previous book resulting from the Second International Symposium on In Situ and On-Site Bioreclamation had no papers on this topic (Means and Hinchee 1994). Recent research publications (Brown et al. 1994; Brown et al. 1995a; Brown et al. 1995b; Dushenkov et al. 1995; Kumar et al. 1995) also demonstrate the interest this research area holds.

Contaminant uptake by plants is often dependent upon the stage of plant growth. Kelley and Guerin (1995) identified the optimal growth period of *Armeria maritima* when the metal content of the aboveground portion was at its highest on a dry weight basis. These results demonstrate that the timing of plant harvest can be critical to maximize the removal of inorganic contaminants from soils.

The advantages of phytoremediation are the low input costs, soil stabilization, the fact that it is aesthetically pleasing (no excavation), and reduced leaching of water and transport of inorganic contaminants in the soil. The main costs are for planting, maintaining plant growth (e.g., fertilizing, watering) harvesting, disposal of contaminated biomass, and repeating the plant-growth cycle. Cornish et al. (1995) give an example where phytoremediation may have three-fold lower costs than soil washing for removing inorganic contaminants from a 0.5 ha waste site. The time frame for estimated cleanup of this site via phytoremediation would be much longer than for soil washing and could take up to 20 years. As the waste site size increases, the costs savings from using phytoremediation versus conventional soil washing should increase. This technology is still in the developmental stage but holds promise for sites with near-surface inorganic contamination.

The limitations of phytoremediation are that the plant must be able to grow in the contaminated soil or material. Sites containing soils with concentrations of inorganic contaminants at inhibitory levels could be amended to dilute the contaminants to concentrations that allow for plant growth. The plant can accumulate only inorganic contaminants that it can reach through root growth or that are soluble in soil water and are transported to the roots via transpiration-driven water movement in the soil. Hence, inorganic contaminants below rooting depth will not be extracted. The process is economical and passive but can take years for contaminant concentrations to reach regulatory levels. Therefore, phytoremediation may not be appropriate for sites with immediate threat or risk to human health or threat of being transported off site. Phytoremediation requires a long-term commitment at the site to ensure adequate plant growth, which may not appeal to clients who require rapid cleanup. A more in-depth discussion of phytoremediation is in the paper by Cunningham et al. (1995) in this book.

SELECTION OF A BIOREMEDIATION STRATEGY FOR INORGANIC CONTAMINANTS

Several questions must be addressed to determine whether bioremediation is a feasible strategy for cleaning up inorganic contaminants at specific site: What is the relative cost of using bioremediation compared to alternative existing technologies? How long will it take for the bioremediation of inorganic contaminants compared to other technologies? Will bioremediation meet the current (and future?) regulatory requirements for the waste type? Are the waste type, concentration, and distribution conducive to bioremediation? Are the site properties such as soil/sediment/water composition, hydrology, presence of other wastes, and factors influencing microbial activity (e.g., pH, Eh, nutrient composition, water, presence of alternative electron acceptors, organic C content, inorganic and organic contaminant toxicity) conducive to bioremediation? Answering these questions usually requires a multidisciplinary approach including expertise from microbiologists, geochemists, hydrologists, and engineers.

Smith and Houthoofd (1995) discuss information and factors to consider when selecting a bioremediation strategy. For mixed contaminants it might be necessary to use multiple treatments or a "treatment train" approach. This treatment train must consider all the contaminants present and use a technology that will not enhance the mobility or toxicity of one contaminant to remove or destroy another. An example provided by Smith and Houthoofd (1995) is the enhancement in the bioremediation of volatile organic compounds by soil vapor extraction. In the process of enhancing biodegradation, Cr(III) will be oxidized to Cr(VI), resulting in increased toxicity and mobility. One advantage of in situ treatment is that ongoing site activities may continue, compared to activities such as excavation that typically require shutdown of that portion of the site. Also, excavation can increase worker exposure to hazardous contaminants.

The papers in this book provide many examples of cases in which bioremediation of inorganic contaminants has been successful. Knowledge of successes as well as failures should make the comparison of bioremediation with other technologies for the remediation of sites contaminated with inorganic compounds more straightforward and thus more attractive to regulators and those responsible for site restoration. Also, fundamental research in the areas of microbe-inorganic interactions, microbial and plant physiology, rhizosphere ecology, biochemistry, geochemistry, hydrology, and engineering will likely yield information to improve concepts discussed in this book as well as yield novel remedial technologies applicable to inorganic contaminants.

ACKNOWLEDGMENT

This research was supported by the Subsurface Science Program, Office of Health and Environmental Research, U.S. Department of Energy (DOE). The continued support of Dr. F. J. Wobber is greatly appreciated. Pacific Northwest Laboratory is operated for the DOE by Battelle Memorial Institute under Contract DE-AC06-76RLO 1830.

REFERENCES

Aickin, R. M., and A.C.R. Dean. 1979. "Lead accumulation by *Pseudomonas fluorescens* and by a *Citrobacter* sp." *Microb. Let. 9*: 55-66.

Aiking, H., K. Kok, H. v. Heerikhuizen, and J. v. T'Riet. 1982. "Adaptation to cadmium by *Klebsiella aerogenes* growing in continuous culture proceeds mainly via formation of cadmium sulfide." *Appl. Environ. Microbiol. 44*: 938-944.

Aiking, H., A. Stijnman, C. v. Garderen, H. v. Heerikhuizen, and J. v. T'Riet. 1984. "Inorganic phosphate accumulation and cadmium detoxification in *Klebsiella aerogenes* NCTC 418 growing in continuous culture." *Appl. Environ. Microbiol. 47*: 374-377.

Anderson, T. A., and J. R. Coats (Eds.) 1994. *Bioremediation Through Rhizosphere Technology*, Vol. 563. American Chemical Society, Washington, DC.

Aronstein, B. N., J. R. Paterek, L. E. Rice, and V. J. Srivastava. 1995. "Effect of chemical pre-treatment on the biodegradation of cyanides." In R. E. Hinchee, J. L. Means, and D. R. Burris (Eds.), *Bioremediation of Inorganics*, pp. 81-87. Battelle Press, Columbus, OH.

Bachofen, R., L. Birch, U. Buchs, P. Ferloni, I. Flynn, G. Jud, H. Tahedi, and T. G. Chasten. 1995. "Volatilization of arsenic compounds by microorganisms." In R. E. Hinchee, J. L. Means, and D. R. Burris (Eds.), *Bioremediation of Inorganics*, pp. 103-108. Battelle Press, Columbus, OH.

Balashova, V. V., and G. A. Zavarzin. 1980. "Anaerobic reduction of ferric iron by hydrogen bacteria." *Microbiology 48*: 635-639.

Barkay, T., R. R. Turner, A. Vandenbrook, and C. Liebert. 1991. "The relationships of Hg(II) volatilization from a freshwater pond to the abundance of *mer* genes in the gene pool of the indigenous microbial community." *Microb. Ecol. 21*: 151-161.

Barnhart, B. J., E. W. Campbell, E. Martinez, D. E. Caldwell, and R. Hallett. 1980. *Potential Microbial Impact on Transuranic Wastes Under Conditions Expected in the Waste Isolation Pilot Plant (WIPP).* National Technical Information Service, Springfield, VA.

Belly, R. T., J. J. Lauff, and C. T. Goodhue. 1975. "Degradation of ethylenediaminetetraacetic acid by microbial populations from an aerated lagoon." *Appl. Environ. Microbiol. 29*: 787-794.

Beveridge, T. J. 1989a. "Interactions of metal ions with components of bacterial cell walls and their biomineralization." In R. K. Poole and G. M. Gadd (Eds.), *Metal Microbe Interactions*, pp. 65-83. OIRL Press, Oxford.

Beveridge, T. J. 1989b. *Metal Ions and Bacteria*. John Wiley and Sons, New York, NY.

Bolton, H., Jr., J. K. Fredrickson, and L. F. Elliott. 1992. "Microbial ecology of the rhizosphere." In F. B. Metting, Jr. (Ed.), *Soil Microbial Ecology*, pp. 27-64. Marcel Dekker, Inc., New York, NY.

Bolton, H., Jr., A. E. Plymale, and D. C. Girvin. 1995. "Influence of aqueous speciation on the biodegradation of EDTA by bacterium BNC1." *Abstracts of the 95th General Meeting of the American Society for Microbiology*: 420.

Brierley, C. L. 1990. "Bioremediation of metal-contaminated surface and groundwaters." *Geomicrobiol. J. 8*: 201-223.

Brown, S. L., R. L. Chaney, J. S. Angle, and A.J.M. Baker. 1994. "Phytoremediation potential of *Thlaspi caerulescens* and bladder campion for zinc- and cadmium-contaminated soil." *J. Environ. Qual. 23*: 1151-1157.

Brown, S. L., R. L. Chaney, J. S. Angle, and A.J.M. Baker. 1995a. "Zinc and cadmium uptake by hyperaccumulator *Thlaspi caerulescens* and metal tolerant *Silene vulgaris* grown on sludge-amended soils." *Environ. Sci. Technol. 29*: 1581-1585.

Brown, S. L., R. L. Chaney, J. S. Angle, and A.J.M. Baker. 1995b. "Zinc and cadmium uptake by hyperaccumulator *Thlaspi caerulescens* grown in nutrient solution." *Soil Sci. Soc. Am. J. 59*: 125-133.

Clarke, S. E., J. Stuart, and J. Sanders-Loehr. 1987. "Induction of siderophore activity in *Anabaena* spp. and its moderation of copper toxicity." *Appl. Environ. Microbiol. 53*: 917-922.

Cornish, J. E., W. C. Goldberg, R. S. Levine, and J. R. Benemann. 1995. "Phytoremediation of soils contaminated with toxic elements and radionuclides." In R. E. Hinchee, J. L. Means, and D. R. Burris (Eds.), *Bioremediation of Inorganics*, pp. 55-63. Battelle Press, Columbus, OH.

Cunningham, S. D., W. R. Berti, and J. W. Huang. 1995. "Remediation of contaminated soils and sludges by green plants." In R. E. Hinchee, J. L. Means, and D. R. Burris (Eds.), *Bioremediation of Inorganics*, pp. 33-54. Battelle Press, Columbus, OH.

Dushenkov, V., P.B.A.N. Kumar, H. Motto, and I. Raskin. 1995. "Rhizofiltration: The use of plants to remove heavy metals from aqueous streams." *Environ. Sci. Technol. 29*: 1239-1245.

Ehrlich, H. L., and C. L. Brierley (Eds.) 1990. *Microbial Mineral Recovery*. McGraw-Hill Publishing Company, New York, NY.

Erardi, F. X., M. L. Failla, and J. O. Falkinham III. 1987. "Plasmid-encoded copper resistance and precipitation by *Mycobacterium scrofulaceum*." *Appl. Environ. Microbiol. 53*: 1951-1954.

Farmer, G. H., D. M. Updegraff, P. M. Radehaus, and E. R. Bates. 1995. "Metal removal and sulfate reduction in low-sulfate mine drainage." In R. E. Hinchee, J. L. Means, and D. R. Burris (Eds.), *Bioremediation of Inorganics*, pp. 17-24. Battelle Press, Columbus, OH.

Ferris, F. G. 1989. "Metallic ion interactions with the outer membrane of gram-negative bacteria." In T. J. Beveridge and R. J. Doyle (Eds.), *Metal Ions and Bacteria*, pp. 295-323. John Wiley & Sons, New York.

Finnegan, I., S. Toerien, L. Abbot, F. Smith, and H. G. Raubenheimer. 1991. "Identification and characterization of an *Acinobacter* sp. capable of assimilation of cyano-metal complexes, free cyanide ions, and organic nitriles." *Appl. Microbiol. Biotech. 36*: 142-144.

Gadd, G. M. 1993. "Microbial formation and transformation of organometallic and organometalloid compounds." *FEMS Microbiol. Rev. 11*: 297-316.

Gadd, G. M., and C. White. 1993. "Microbial treatment of metal pollution — A working biotechnology." *TIBTech 11*: 353-359.

Garbisu, C., T. Ishii, N. R. Smith, B. C. Yee, D. L. Carlson, A. Yee, B. B. Buchanan, and T. Leighton. 1995. "Mechanisms regulating the reduction of selenite by aerobic gram (+) and (-) bacteria." In R. E. Hinchee, J. L. Means, and D. R. Burris (Eds.), *Bioremediation of Inorganics*, pp. 125-131. Battelle Press, Columbus, OH.

Geesey, G. G., and L. Jang. 1989. "Interactions between metal ions and capsular polymers." In T. J. Beveridge and R. J. Doyle (Eds.), *Metal Ions and Bacteria*, pp. 325-357. John Wiley & Sons, New York, NY.

Graydon, J. W., S. M. Oakley, B. H. Reed, and H. L. Ball. 1995. "On-site biological nitrogen removal using recirculating trickling filters." In R. E. Hinchee, J. L. Means, and D. R. Burris (Eds.), *Bioremediation of Inorganics*, pp. 133-139. Battelle Press, Columbus, OH.

Harwood-Sears, V., and A. S. Gordon. 1990. "Copper-induced production of copper-binding supernatant proteins by the marine bacterium *Vibrio alginolyticus*." *Appl. Environ. Microbiol. 56*: 1327-1332.

Higham, D. P., P. J. Sadler, and M. D. Scawen. 1984. "Cadmium-resistant *Pseudomonas putida* synthesizes novel cadmium proteins." *Science 225*: 1043-1046.

Higham, D. P., P. J. Sadler, and M. D. Scawen. 1985. "Cadmium-resistant *Pseudomonas putida*: Growth and uptake of cadmium." *J. Gen. Microbiol. 131*: 2539-2544.

Higham, D. P., P. J. Sadler, and M. D. Scawen. 1986a. "Cadmium-binding proteins in *Pseudomonas putida*: Pseudothioneins." *Environ. Health Perspec. 65*: 5-11.

Higham, D. P., P. J. Sadler, and M. D. Scawen. 1986b. "Effect of cadmium on the morphology, membrane integrity and permeability of *Pseudomonas putida*." *J. Gen. Microbiol. 132*: 1475-1482.

Kelley, R. J., and T. F. Guerin. 1995. "Feasibility of using hyperaccumulating plants to bioremediate metal-contaminated soil." In R. E. Hinchee, J. L. Means, and D. R. Burris (Eds.), *Bioremediation of Inorganics*, pp. 25-32. Battelle Press, Columbus, OH.

Khazaeli, M. B., and R. S. Mitra. 1981. "Cadmium-binding component in *Escherichia coli* during accommodation to low levels of this ion." *Appl. Environ. Microbiol. 41*: 46-50.

Kumar, P.B.A.N., V. Dushenkov, H. Motto, and I. Raskin. 1995. "Phytoextraction: The use of plants to remove heavy metals from soils." *Environ. Sci. Technol. 29*: 1232-1238.

Lauff, J. J., D. B. Steele, L. A. Coogan, and J. M. Breitfeller. 1990. "Degradation of the ferric chelate of EDTA by a pure culture of an *Agrobacterium* sp." *Appl. Environ. Microbiol. 56*: 3346-3353.

Lewin, R. 1984. "How microorganisms transport iron." *Science 225*: 401-402.

Lortie, L., W. D. Gould, S. Rajan, R.G.L. McCready, and K. J. Cheng. 1992. "Reduction of selenate and selenite to elemental selenium by a *Pseudomonas stutzeri* isolate." *Appl. Environ. Microbiol. 58*: 4042-4044.

Lovley, D. R. 1994. "Microbial reduction of iron, manganese, and other metals." *Adv. Agron. 54*: 175-231.

Lu, Y., and E. Wilkins. 1995. "Heavy metal removal by caustic-treated yeast immobilized in alginate." In R. E. Hinchee, J. L. Means, and D. R. Burris (Eds.), *Bioremediation of Inorganics*, pp. 117-124. Battelle Press, Columbus, OH.

Macaskie, L. E. 1990. "An immobilized cell bioprocess for the removal of heavy metals from aqueous flows." *J. Chem. Technol. Biotech. 49*: 357-364.

Macaskie, L. E., and A.C.R. Dean. 1984. "Cadmium accumulation by a *Citrobacter* sp." *J. Gen. Microbiol. 130*: 53-62.

Macaskie, L. E., A.C.R. Dean, A. K. Cheetham, R.J.B. Jakemen, and A. J. Skarnulis. 1987. "Cadmium accumulation by a *Citrobacter* sp.: The chemical nature of the accumulated metal precipitate and its location on the bacterial cells." *J. Gen. Microbiol. 133*: 539-544.

Macaskie, L. E., R. M. Empson, A. K. Cheetham, C. P. Grey, and A. J. Skarnulis. 1992. "Uranium bioaccumulation by a *Citrobacter* sp. as a result of enzymatically mediated growth of polycrystalline HUO_2PO_4." *Science 257*: 782-784.

Macy, J. M., T. A. Michel, and D. G. Kirsch. 1989. "Selenate reduction by a *Pseudomonas* species: A new mode of anaerobic respiration." *FEMS Microb. Lett. 61*: 195-198.

McBride, B. C., and R. S. Wolfe. 1971. "Biosynthesis of dimethylarsine by *Methanobacterium*." *Biochemistry 10*: 312-317.

McCarty, S. L., T. G. Chasteen, V. Stalder, and R. Bachofen. 1995. "Bacterial bioremediation of selenium oxyanions using a dynamic flow bioreactor and headspace analysis." In R. E. Hinchee, J. L. Means, and D. R. Burris (Eds.), *Bioremediation of Inorganics*, pp. 95-102. Battelle Press, Columbus, OH.

Means, J. L., and R. E. Hinchee (Eds.). 1994. *Emerging Technology for Bioremediation of Metals*. 148 pp. Lewis Publishers, Boca Raton, FL.

Mitra, R. S., R. H. Gray, B. Chin, and I. A. Bernstein. 1975. "Molecular mechanisms of accommodation in *Escherichia coli* to toxic levels of Cd^{2+}." *J. Bacteriol. 121*: 1180-1188.
Mittelman, M. W., and G. G. Geesey. 1985. "Copper-binding characteristics of exopolymers from a freshwater-sediment bacterium." *Appl. Environ. Microbiol. 49*: 846-851.
Nagle, N. J., C. J. Rivard, A. Mohangheghi, and G. P. Philippidis. 1995. "Bioconversion of cyanide and acetonitrile by a municipal-sewage-derived anaerobic consortium." In R. E. Hinchee, J. L. Means, and D. R. Burris (Eds.), *Bioremediation of Inorganics*, pp. 71-79. Battelle Press, Columbus, OH.
Nörtmann, B. 1992. "Total degradation of EDTA by mixed cultures and a bacterial isolate." *Appl. Environ. Microbiol. 58*: 671-676.
Obuekwe, C. O., and D.W.S. Westlake. 1982. "Effects of medium composition on cell pigmentation, cytochrome content, and ferric iron reduction in a *Pseudomonas* sp. isolated from crude oil." *J. Microbiol. 28*: 989-992.
Owens, L. P., K. C. Kovac, J. L. Kipps, and D.W.J. Hayes. 1995. "Biological reduction of soluble selenium in subsurface agricultural drainage water." In R. E. Hinchee, J. L. Means, and D. R. Burris (Eds.), *Bioremediation of Inorganics*, pp. 89-94. Battelle Press, Columbus, OH.
Poole, R. K., and G. M. Gadd (Eds.) 1989. *Metal-Microbe Interactions*. OIRL Press at Oxford University Press, Oxford, UK.
Premuzic, E. T., M. Lin, A. J. Francis, and J. Schubert. 1986. "Production of chelating agents by *Pseudomonas aeruginosa* grown in the presence of thorium and uranium." In R. A. Bulman and J. R. Cooper (Eds.), *Speciation of Fission and Activation Products in the Environment*, pp. 391-397. Elsevier, New York, NY.
Robinson, J. B., and O. H. Touvinen. 1984. "Mechanisms of microbial resistance and detoxification of mercury and organomercury compounds: Physiological, biochemical, and genetic analyses." *Microbiol. Rev. 48*: 95-124.
Rusin, P. A., L. Quintana, J. R. Brainard, B. A. Strietelmeier, C. D. Tait, S. A. Ekberg, P. D. Palmer, T. W. Newton, and D. L. Clark. 1994. "Solubilization of plutonium hydrous oxide by iron-reducing bacteria." *Environ. Sci. Technol. 28*: 1686-1690.
Saouter, E., R. Turner, and T. Barkay. 1994. "Mercury microbial transformations and their potential for the remediation of a mercury-contaminated site." In J. L. Means and R. E. Hinchee (Eds.), *Emerging Technology for Bioremediation of Metals*, pp. 99-104. Lewis Publishers, Boca Raton, FL.
Schmidt, G. C., and M. B. Ballew. 1995. "In situ biodenitrification of nitrate surface water." In R. E. Hinchee, J. L. Means, and D. R. Burris (Eds.), *Bioremediation of Inorganics*, pp. 109-116. Battelle Press, Columbus, OH.
Schreiber, D. R., F. J. Millero, and A. S. Gordon. 1990. "Production of an extracellular copper-binding compound by the heterotrophic marine bacterium *Vibrio alginolyticus*." *Marine Chem. 28*: 275-284.
Silva-Avalos, J., M. G. Richmond, O. Nagappan, and D. A. Kunz. 1990. "Degradation of metal-cyano complex tetracyanonickelate(II) by cyanide-utilizing bacterial isolates." *Appl. Environ. Microbiol. 56*: 3664-3670.
Silver, S., and M. Walderhaug. 1992. "Gene regulation of plasmid- and chromosome-determined inorganic ion transport in bacteria." *Microbiol. Rev. 56*: 195-228.
Smith, L. A., and J. M. Houthoofd. 1995. "Considerations in deciding to treat contaminated soils in situ." In R. E. Hinchee, J. L. Means, and D. R. Burris (Eds.), *Bioremediation of Inorganics*, pp. 149-164. Battelle Press, Columbus, OH.
Summers, A. O., and T. Barkay. 1989. "Metal resistance genes in the environment." In S. Levy and R. Miller (Eds.), *Gene Transfer in the Environment*, pp. 287-308. McGraw Hill, New York, NY.
Tuovinen, O. H. 1990. "Biological fundamentals of mineral leaching processes." In H. L. Ehrlich and C. L. Brierley (Eds.), *Microbial Mineral Recovery*, pp. 55-77. McGraw-Hill Publishing Company, New York, NY.

Volesky, B. (Ed.) 1990a. *Biosorption of Heavy Metals.* CRC Press, Boca Raton, FL.

Volesky, B. 1990b. "Removal and recovery of heavy metals by biosorption." In B. Volesky (Ed.), *Biosorption of Heavy Metals*, pp. 7-43. CRC Press, Boca Raton, FL.

Wang, T., J. C. Weissman, G. Ramesh, R. Varadarajan, and J. R. Benemann. 1995. "Bioremoval of toxic elements with aquatic plants and algae." In R. E. Hinchee, J. L. Means, and D. R. Burris (Eds.), *Bioremediation of Inorganics*, pp. 65-69. Battelle Press, Columbus, OH.

Whitlock, J. L. 1990. "Biological detoxification of precious metal processing wastewaters." *Geomicrobiol. J. 8*: 241-249.

Wilber, C. G. 1980. "Toxicity of selenium: A review." *Clin. Toxicol. 17*: 171-230.

Wildeman, T., J. Gusek, J. Cevaal, K. Whiting, and J. Scheuering. 1995. "Biotreatment of acid rock drainage at a gold-mining operation." In R. E. Hinchee, J. L. Means, and D. R. Burris (Eds.), *Bioremediation of Inorganics*, pp. 141-148. Battelle Press, Columbus, OH.

Wildung, R. E., and T. R. Garland. 1980. "The relationship of microbial processes to the fate and behavior of transuranic elements in soils, plants, and animals." In W. C. Hanson (Ed.), *Transuranic Elements in the Environment*, Vol. NVO-178, pp. 127-169. National Technical Information Service, Springfield, VA.

Wildung, R. E., T. R. Garland, and H. Drucker. 1979. "Nickel complexes with soil microbial metabolites-mobility and speciation in soils." In E. A. Jenne (Ed.), *Chemical Modeling in Aqueous Systems Speciation, Sorption, Solubility, and Kinetics*, pp. 181-200. American Chemical Society, Washington, DC.

Metal Removal and Sulfate Reduction in Low-Sulfate Mine Drainage

Garry H. Farmer, David M. Updegraff, Petra M. Radehaus, and Edward R. Bates

ABSTRACT

A treatability study using two continuous upflow bioreactors was conducted to evaluate the potential removal of metal contamination, primarily zinc, from mine drainage with constructed wetlands that incorporate sulfate-reducing bacteria (SRB). The drainage from the Burleigh Tunnel in Silver Plume, Colorado, contains low levels of sulfate (350 to 550 mg/L) that may limit the production of hydrogen sulfide by sulfate-reducing bacteria, thus limiting metal removal by the system. Total metals, anions, and field parameters in the mine drainage and the bioreactor effluents were routinely analyzed over 8 weeks. In addition, the bioreactor compost packing was analyzed for metals and sulfate-reducing bacteria. Zinc removal in both reactors was in excess of 99% (average influent zinc concentration of 54.4 mg/L) after 8 weeks of operation. Furthermore, sulfate-reducing bacteria in the bioreactor compost ranged from 10^5 to 10^6 colony-forming units (CFUs) per gram of compost.

INTRODUCTION

The contamination of streams, rivers, and lakes with metals originating from mining activities has become a serious environmental problem in many areas of the United States. In Colorado, the Colorado Department of Public Health and Environment estimates that 1,300 mi (2,092 km) of streams and rivers have been affected by mine drainage (USGS 1994). Over the past 10 to 15 years, numerous researchers have applied constructed wetlands technology to the remediation of mine drainage. Currently, constructed wetlands appears to be one of the few cost-effective treatments available for the remediation of mine drainage. Thus, the U.S. Environmental Protection Agency (U.S. EPA) is evaluating the constructed wetlands technology at the Burleigh Tunnel (Silver Plume, Colorado) within the Superfund Innovative Technology Evaluation (SITE) program.

In general, there are two types of constructed wetlands: free water systems (FWS) and subsurface flow systems (SFS). FWS typically are composed of shallow channels or ponds and remove metals by chemical oxidation followed by precipitation of the metal oxide. SFS channel the mine drainage through a porous material with a high organic content such as compost. SRB within the compost produce hydrogen sulfide that reacts with the dissolved metals to form insoluble or slightly soluble metal sulfides. The metal sulfides precipitate and are filtered from the water by the compost.

The purpose of this study was to evaluate the potential of SRB to remove metal contamination from mine drainage containing a moderately low concentration of sulfate (350 to 550 mg/L) with compost-filled bioreactors. In addition, the study evaluated several parameters indicating the level of sulfate-reduction activity including sulfate depletion, sulfide generation, and sulfate-reducing bacteria populations.

EXPERIMENTAL PROCEDURES AND MATERIALS

Two continuous upflow bioreactors were set up at the Burleigh Tunnel during June 1993. The reactor packing (substrate) consisted of a composted mixture of dairy cow manure and paper products (96% by weight) and alfalfa hay (4%). The smaller bioreactor was approximately 4 ft (1.2 m) tall and 22 in. (56 cm) in diameter and held approximately 60 gal (230 L) of substrate and water. The larger reactor was approximately 8 ft (2.4 m) tall and 22 in. (56 cm) in diameter and held approximately 130 gal (490 L) of substrate and water. The lower 6 to 12 in (15 to 30 cm) of each bioreactor was filled with gravel to support inlet piping and minimize channeling.

Peristaltic pumps removed the mine drainage from a small sump located about 100 ft (30 m) from the Burleigh Tunnel entrance. Flowrates for the small reactor ranged from 20 to 30 mL/min, to 50 to 60 mL/min for the larger reactor. The flowrates were established to provide a residence time between 50 and 100 h.

Influent water samples were collected from the sump receiving the mine drainage. Bioreactor effluent samples were collected from a joint in the outflow tubing on each reactor directly into the sample containers. Substrate samples were collected with a 1-L polyethylene dipper with a 3-ft-long (1-m-long) handle. The dipper was inserted into the substrate as far as possible, typically between 2 and 2.5 ft (0.62 and 0.77 m).

All influent and effluent water samples and substrate samples were analyzed for total metals by inductively coupled plasma emission spectroscopy (ICP) or inductively coupled plasma mass spectrometry (ICPMS) using U.S. EPA protocols. Anions were also analyzed by U.S. EPA protocols using gravimetric and spectroscopic techniques for alkalinity, sulfate, nitrate, nitrite, chloride, fluoride, and sulfide (effluent only). Field measurements included the determination of pH, Eh, dissolved oxygen, and conductivity in influent and effluent water.

In addition, substrate samples were analyzed for sulfate-reducing bacteria by a direct counting procedure using serial dilutions of the substrate sample into

10-mL-deep test tubes containing a lactate medium (Modified Media E, Postgate 1984). Finally, enrichment cultures of SRB were prepared from deep tube cultures by removing individual black colonies from the medium with a sterile pipette and transferring them to fresh lactate medium.

RESULTS

Table 1 contains influent and effluent sample results from both bioreactors. The influent analyses indicate that the primary metals contained in the Burleigh drainage are calcium (75 to 100 mg/L), magnesium (38 to 51 mg/L), zinc (43 to 69 mg/L), sodium (2.7 to 16 mg/L), potassium (2.5 to 4.5 mg/L), and manganese (1.8 to 3.6 mg/L). The remaining metals analyzed during the study (aluminum, arsenic, cadmium, iron, lead, nickel, and silver) were detected in concentrations less than 0.5 mg/L in influent samples.

Initially, the small reactor effluent samples contained elevated levels of potassium (2,610 mg/L), sodium (471 mg/L), magnesium (364 mg/L), and calcium (235 mg/L) that decreased over the 8 weeks of the study. Zinc in the small reactor effluent ranged from 0.6 to 9.4 mg/L, manganese from 0.18 to 0.82 mg/L, and nickel from 0.01 to 0.07 mg/L.

The large reactor effluent also contained elevated levels of potassium (3,210 mg/L), sodium (560 mg/L), magnesium (383 mg/L), and calcium (237 mg/L) at startup that decreased throughout the study. Zinc in the effluent of the large reactor ranged from 0.42 to 2.4 mg/L, manganese from 0.17 to 0.73 mg/L, and nickel from 0.084 to 0.014 mg/L.

Elevated levels of chloride, sulfate, phosphorous, and ammonia also were present in the effluent of each reactor at startup; however, their concentrations decreased substantially after 2 weeks of operation. Initially, the concentration of sulfate in the small reactor effluent was high (1,220 mg/L), but rapidly dropped to a low of 9.7 mg/L, followed by a gradual increase to between 170 to 180 mg/L. Sulfate concentrations in the large reactor were high at startup (922 mg/L), but dropped rapidly to nondetect levels (5 mg/L), and then increased to a somewhat stable level between days 20 and 46. Sulfide levels in the effluents of each reactor slowly increased from nondetect (reporting limit 1.2 mg/L) to a maximum of 161 mg/L in the small reactor and 34.1 mg/L in the large reactor. However, the sulfide data of both systems are somewhat scattered.

Two substrate samples were collected from each bioreactor and analyzed during the treatability study. Samples collected at the beginning of the study indicated that the substrate contained approximately 150 mg/kg of zinc, 60 to 70 mg/kg of manganese, and 0.20 to 0.25 mg/kg of cadmium. In addition, the substrate contained elevated levels of aluminum (1,200 mg/kg), cadmium (5,100 mg/kg), iron (2,000 mg/kg), potassium (4,000 mg/kg), and sodium (550 mg/kg). The results of metals analysis conducted on substrate samples collected on day 35 indicated that manganese was accumulating in the substrate material; however, zinc, cadmium, nickel, and other metals were not present in higher concentrations compared to initial substrate sample results.

TABLE 1. Bioreactor influent and effluent mean monthly results (where NS is not sampled and ND is not detected).

Analyte	Report Limit	Influent			Small Reactor Effluent			Larger Reactor Effluent		
		June	July	August	June	July	August	June	July	August
Metals										
Aluminum	0.1	0.24	0.058	0.036	0.36	0.53	0.24	0.46	0.21	0.38
Arsenic	0.0010	0.0023	0.0052	ND	0.030	0.011	0.0041	0.031	0.0099	0.0068
Cadmium	0.00030	0.094	0.12	0.12	0.21	0.014	0.0019	0.0050	0.0038	0.0015
Calcium	0.2	86	93	90	170	120	120	160	120	140
Iron	0.1	0.41	0.29	0.21	9.1	3.0	0.59	11	3.7	0.90
Lead	0.0010	0.030	0.026	0.011	0.056	0.026	0.009	0.042	0.011	0.012
Magnesium	0.2	46	48	46	230	120	80	250	140	100
Manganese	0.01	1.8	2.4	2.2	0.46	0.32	0.20	0.52	0.32	0.20
Nickel	0.04	0.043	0.046	0.047	0.045	0.018	0.012	0.055	0.062	0.014
Potassium	5	3.1	3.3	2.8	1500	290	90	2000	340	140
Silver	0.0001	0.00014	0.0008	0.00031	0.00037	0.00027	0.00013	0.00033	0.00012	0.00014
Sodium	5	15	6.9	14	270	63	20	340	57	20
Zinc	0.02	49	57	55	5.4	4.6	0.94	2.0	1.3	0.52

TABLE 1. (continued).

Analyte	Report Limit	Influent			Small Reactor Effluent			Larger Reactor Effluent		
		June	July	August	June	July	August	June	July	August
Anions										
Sulfate	2.5	390	440	430	520	140	180	460	94	130
Sulfide, Total	0.05	NS	NS	NS	1.6	6.9	14	1.6	3.5	26
Fluoride	0.10	0.95	0.89	1.1	0.23	0.28	0.45	0.16	0.15	0.22
Chloride	0.5	14	16	17	770	110	28	880	140	27
Phosphorous, Total	0.05	ND	ND	ND	320	230	40	370	250	48
Orthophosphate	0.05	ND	ND	ND	190	120	43	310	140	49
Nitrite as N	0.01	ND	ND	ND	0.22	ND	ND	0.30	ND	ND
Nitrate plus Nitrite	0.1	ND	ND	ND	ND	ND	ND	ND	ND	ND
Nitrate as N	0.1	ND	ND	ND	0.22	ND	ND	0.30	ND	ND
Ammonia	0.1	ND	ND	ND	46	24	15	85	25	16
Total Solids										
TSS	2.0	19	17	13	53	73	28	100	36	39
TDS	10.0	720	790	820	3900	2200	1100	10000	2600	1400
TOC	50	NS	NS	NS	NS	NS	NS	NS	NS	NS
Alkalinity										
As $CaCO_3$	5.0	110	110	110	1900	970	600	2600	1100	770
Field Results										
pH (pH units)		6.7	7.2	7.0	6.4	6.7	7.1	6.8	6.7	7.0
T (C)		14	10	12.4	15	15	14	16	16	12
Conductivity (µS)		970	980	1500	4800	2200	1400	1000	1600	1500
Eh (mV)		NS	NS	NS	–80	–130	–320	–80	–120	–320
Dissolved Oxygen (mg/L)		8.0	8.0	8.0	NS	NS	NS	NS	NS	NS

NS = not sampled; ND = not detected.

DISCUSSION

Previous work with SFS wetlands and bioreactors containing SRB (Eger et al. 1993; Hammack et al. 1994; Fyson et al. 1994; and Wildeman et al. 1992) has found that arsenic, cadmium, copper, lead, iron, nickel, and zinc are removed as sulfides or coprecipitate with sulfide precipitation. The comparison of effluent to influent concentrations (Table 1) during this study indicates that zinc, arsenic, cadmium, manganese, nickel, and silver were removed by the compost and hay reactors. The removal of arsenic, cadmium, and nickel is significant because these metals are present in low concentrations in the Burleigh drainage.

Figure 1 shows the percent removal of zinc in both reactors over the 8 weeks of the study. The pattern observed for the small reactor, a high initial removal followed by a steep drop in removal efficiency, followed by a gradual increase of zinc removal, was observed in constructed wetlands studies conducted by Machemer and Wildeman (1992). They suggest that the initial high removal phase results from sorption of the metals to the compost; once sorption sites are filled, removal efficiency drops. Gradually, SRB become established and metal removal reflects the rate of sulfate reduction.

Manganese was consistently removed from the mine drainage by both reactors (Figure 2). Removal of manganese by the small reactor ranged from 73% to 92% between day 5 and day 56 with an average removal efficiency of 84%.

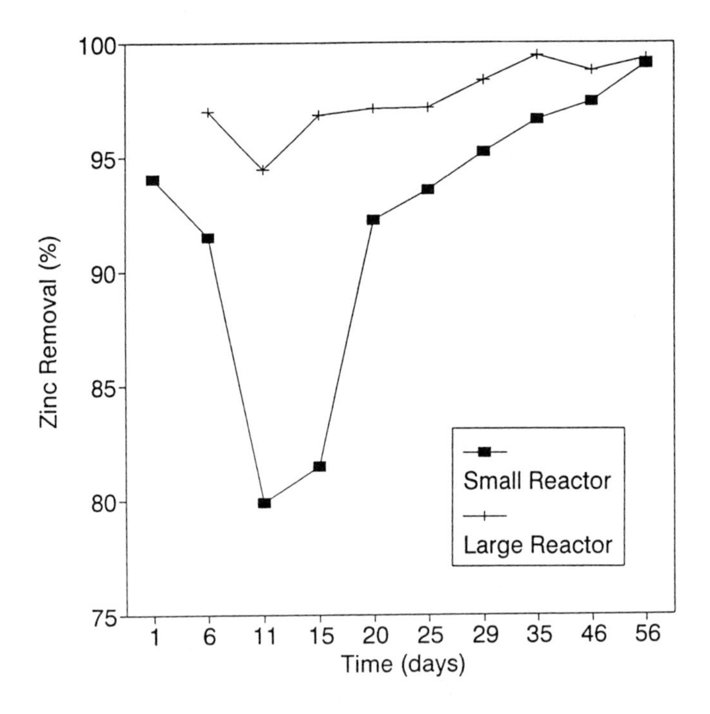

FIGURE 1. Graph of zinc removal in bioreactors.

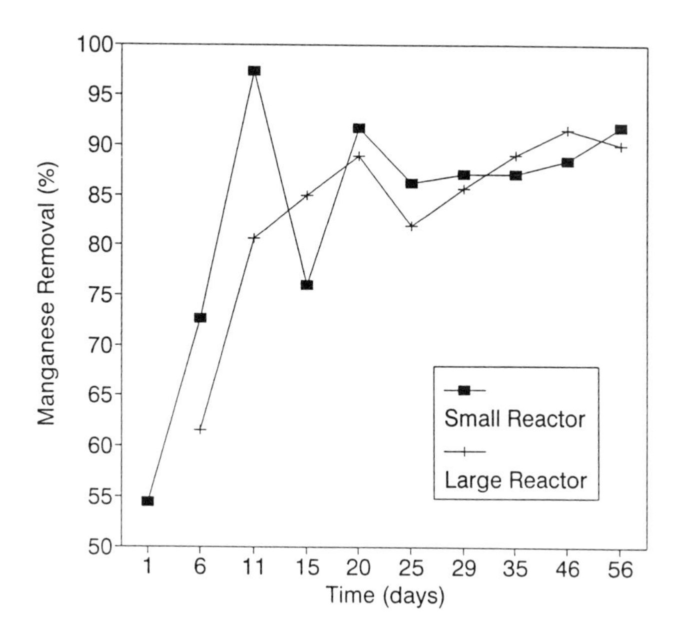

FIGURE 2. Graph of manganese removal in bioreactors.

The large reactor showed a similar performance, removing an average of 87% of the manganese between day 5 and day 50. Manganese is not expected to form a sulfide at the pH or Eh conditions in the reactors; however, there is sufficient carbonate in these systems that $MnCO_3$ (rhodochrosite) may precipitate.

Figure 3 shows the percent removal of cadmium in both reactors over the 8 weeks of the study. The observed patterns for cadmium removal are similar to those observed for zinc. Initially, cadmium removal in the small reactor was about 87% this level dropped to 70%, followed by a gradual increase to a final removal efficiency exceeding 97%. Cadmium removal in the large reactor started at 95% and increased to greater than 99% over the 8 weeks of the study.

Results of serial dilutions of substrate samples collected on day 25 found sulfate-reducing bacteria present in excess of 10^6 CFUs per gram of substrate in both reactors. Further, enrichment cultures of the SRB present in the deep tubes were observed by phase-contrast light microscopy. Based on morphology alone, these bacteria are believed to be several species of *Desulfovibrio* and one of *Desulfobacter*. No attempt was made to further identify these sulfate-reducing bacteria.

Sulfate concentrations in the small reactor over the final 3 weeks of the study suggest that dissimilatory sulfate-reduction activity had stabilized and that the bacteria were converting an average 254 mg/L of the 429 mg/L available sulfate, or roughly 60%. Sulfate use in the large reactor between days 29 and 46 was 330 mg/L, or 78% of the available sulfate. These results suggest that sulfate use is a satisfactory indicator of sulfate-reducing activity and that a 1- or 2-m-thick constructed wetland system at this location will not be sulfate limited.

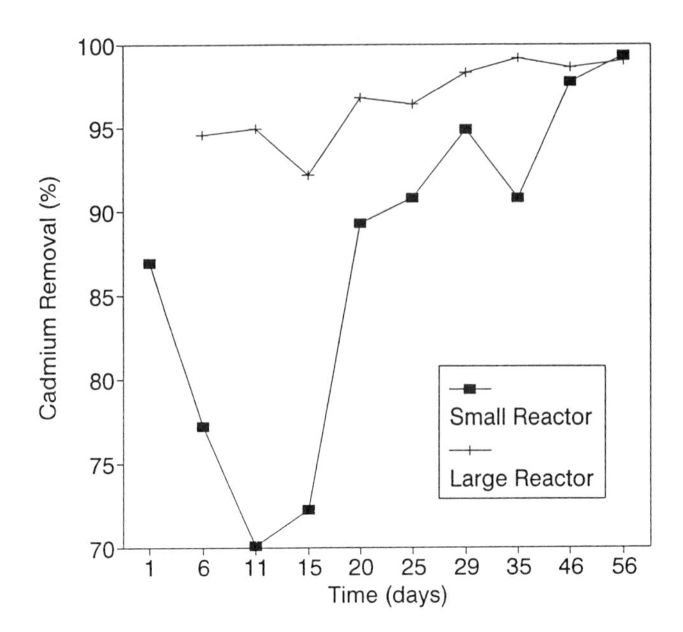

FIGURE 3. Graph of cadmium removal in bioreactors.

REFERENCES

Eger, P., G. Melchert, D. Antonson, and J. Wagner. 1993. "The Use of Wetland Treatment to Remove Trace Metals From Mine Drainage." In G. A. Moshiri (Ed.), *Constructed Wetlands For Water Quality Improvement*. pp. 171-178. Lewis Publishers, Boca Raton, FL.

Fyson, A., M. Kalin, and L. W. Adrian. 1994. "Arsenic and Nickel Removal by Wetland Sediments." In *Proceedings of International Land Reclamation and Mine Drainage Conference*. United States Department of Interior Bureau of Mines Special Publication SP 06A-94. pp. 109-118.

Hammack, R. W., D. H. Dvorack, and H. M. Edenborn. 1994. "Bench-Scale Test to Selectively Recover Metals From Metal Mine Drainage Using Biogenic H_2S." In *Proceedings of International Land Reclamation and Mine Drainage Conference*. United States Department of Interior Bureau of Mines Special Publication SP 06A-94. pp. 214-222.

Machemer, S. D., and T. R. Wildeman. 1992. "Organic Complexation Compared with Sulfide Precipitation as Metal Removal Processes from Acid Mine Drainage in a Constructed Wetland." *Journal of Contaminant Hydrology*, *9*:115-131.

Postgate, J. R. 1984. *The Sulfate-Reducing Bacteria*, 2nd ed. Cambridge University Press, London, England.

USGS, United States Geological Survey. 1994. *Guidebook on the Geology, History, and Surface-Water Contamination and Remediation in the Area From Denver to Idaho Springs, Colorado.* USGS Circular 1097.

Wildeman, T. R., G. A. Brodie, and J. J. Gusek. 1992. *Wetland Design for Mining Operations.* Bitek Publishers, Vancouver, BC.

Feasibility of Using Hyperaccumulating Plants to Bioremediate Metal-Contaminated Soil

Robert J. Kelly and Turlough F. Guerin

ABSTRACT

A feasibility study was carried out to determine whether selected plants were capable of hyperaccumulating anthropogenic sources of metals found in soils from three contaminated sites. A trial was conducted using the previously reported hyperaccumulators, *Armeria maritima* (thrift), *Impatiens balsamina* (balsam), *Alyssum saxatile* (gold dust), and the control species, *Brassica oleracea* (cabbage). Although none of these plants showed any substantial hyperaccumulation of Cu, Zn, Pb, and Cd, it was established that there is an optimum period in the life-cycle of these plants in which the metal concentration reaches a maximum. This period was dependent on the metal, soil, and plant type. The current paper describes the data obtained for Zn and Cu uptake by thrift.

INTRODUCTION

Decontamination of soils containing heavy metals remains one of the most intractable problems for bioremediation. Current remediation techniques are based on either immobilization, extraction by physico-chemical methods, or burial. These techniques often require special equipment and operators, are expensive, can remove biological activity from the soil, and can deleteriously affect the soil physical properties (McGrath et al. 1994). In previous research, considerable attention has been given to the area of plant uptake of inorganic compounds; however, relatively little practical consideration has been given to applying the process of metal uptake and storage for the purpose of cleaning up contaminated soil. McGrath et al. (1994) have indicated through their research that cropping with metal-accumulating plants may be a means of removing metals from metal-contaminated soils.

It is generally recognized that a number of constraints need to be overcome before the concept of metal-hyperaccumulators will become a technology for the remediation of metal-contaminated soil. These constraints include the type of accumulation mechanism in the plant, the choice of plants for a contaminated site, the soil type, and the origin of the metal. However, the most important constraint is probably that of the length of time required for the plant to translocate and reduce the soil metal concentrations to acceptable levels. The current study aimed to determine the period in the life cycle of a previously reported hyperaccumulator at which the metal concentration is at its greatest. The current study also aimed to determine the effect of metal uptake from three soils from typical metal-contaminated sites.

EXPERIMENTAL PROCEDURES AND MATERIALS

Selection of Soils and Plants

Soil was obtained from three contaminated sites in Sydney, New South Wales. The soil types chosen were different, allowing comparisons to be made between the availability of the metals in the different soil types. Soil A was known to contain high levels of Pb; Soil B was obtained from an old tannery and was found to contain As, Zn, Cu, and Pb; and Soil C was known to contain As, Cd, Cu, Hg, Pb, and Zn. In the current trial only Cu, Zn, Pb, and Cd were analyzed. Soil pH, particle size analyses, electrical conductivity (EC) of soil water extract, organic matter, and metals were determined by standard U.S. Environmental Protection Agency (EPA) techniques. These are reported in detail in Kelly (1994).

The plant species were selected on their reported ability to accumulate and/or hyperaccumulate heavy metals and their availability. Based on availability and previous research, the following plants were chosen: (1) *Armeria maritima*, thrift (Plumbaginaceae); (2) *Impatiens balsamina*, balsam (Balsaminaceae); (3) *Alyssum saxatile*, gold dust (Crucifereae); and (4) *Brassica oleracea*, cabbage (Crucifereae). Of these plants *A. maritima* (thrift) has been classified as a hyperaccumulator of Pb (Baker and Brooks 1989) and has been found growing on metalliferous soils with as high as 1% Zn. It is also reported to be an indicator for Cu. *I. balsamina* is an indicator plant for Cu, although it has not been previously assessed for hyperaccumulating other metals. *A. saxatile* (gold dust) belongs to a genus of hyperaccumulators, but little phytoremediation research has been conducted on this particular species. *B. oleracea* was used as a control because it is not a hyperaccumulator but is known to be tolerant of high metal concentrations.

Design of Trial

The 108 plant pots had soil added to them with moisture content at one-half field capacity (FC) and packed to give a bulk density of 1.3 g/cm^3. After this the soils were made up to near FC, and the pregerminated seeds were sown. A basal fertilizer application was added to each pot, and polythene beads were placed on the top of the soil to minimize evaporation (Kelly 1994). The pots were kept at FC throughout the trial by regular weighing (Kelly 1994). The pots were placed in a constant-temperature water bath at 20°C. The plants were harvested at three time intervals. The whole plant was harvested, and the root zone was washed free of soil. After oven drying (70°C), the plants were analyzed for the metals. Subsamples of soil from around the root zone of each pot were taken immediately after harvesting the plants. This soil was used to determine the total heavy metal content. The design of the trial was a split plot time design based on a randomized complete block design. An analysis of variance was carried out to establish if there was a difference in the amounts of metals accumulated by each plant on the three different soil types.

RESULTS AND DISCUSSION

The three soils used were from different textural classes. This was expected to affect the way in which the plants accumulated the metals. Soil A was a clay loam containing 34% clay, whereas Soil B was a clay containing 45% clay. Because soils A and B had such high clay content, metal adsorption to the clay surface was expected to be higher than that of Soil C which contained only 11% clay. However, the high organic matter in Soil C was also likely to result in a reduction in the amount of metals available to the plants. The organic matter was very high in Soil C (14.1%) compared to A (1.8%) and B (2.2%). This was due to the way in which the soil became contaminated. Incinerated material (ash) was mixed with sand and dumped on site, and this led to a high organic matter content.

The concentrations of metals in the three soils are presented in Table 1. The amount of available metals as a percentage of total metal content is reflected to some extent by the soil type and organic matter content. In general, for a clay soil the available metal content ranges from 30 to 80% and for a sandy soil 50 to 99% (Alloway 1990 cited in Kelly 1994).

It was found that the plants growing on Soil C, a loamy sand, accumulated considerably less metal than those growing on Soils A and B. It was expected that the accumulation would have been greater on Soil C due to the fact that this type of soil would have more bioavailable metals. This result can be explained by the large amount of organic matter in Soil C complexing the metals and thus reducing availability to the plant.

A number of factors affect the plant uptake of metals including soil type, amount of available metal, and competition from other metals and ions. The

TABLE 1. Total metal concentrations in soil used in the trial.

	Metal Concentrations (mg/kg)			
Soil	**Cu**	**Zn**	**Pb**	**Cd**
A	30	182	167	1.0
B	44	248	56	1.1
C	257	1,765	1,274	3.0

data for the uptake of Zn and Cu by thrift from the three soils are presented in Figures 1 and 2, respectively. The difference in plant tissue concentrations in these graphs is primarily explained by the different initial soil metal concentrations. Of the plants studied, thrift was the only plant that has previously been classified as a true hyperaccumulator (Baker and Brooks 1989). Thrift was reported to hyperaccumulate lead and has also been reported to be an indicator of Cu (Baker 1981). Brooks (1994) has quantitatively defined indicators of copper by giving them a specific indicator value (s.i.v.):

$$\text{s.i.v.} = \frac{\text{Highest Cu Concentration in Individual Plant Sampled} - \text{Lowest Cu Concentration in Individual Plant Sampled}}{\text{Total Soil Cu Concentration}}$$

The Cu concentrations in this equation refer to those in the individual plants sampled from the total population of plants taken from the topsoil overlying the mineral deposit or soil contamination. It has been suggested that good indicators should have an s.i.v. of 4 or less. The values obtained from the current research were 1.3 for Soil A, 0.29 for Soil B, and 0.21 for Soil C. Based on the above criteria, these values would classify thrift as an indicator plant for Cu. Table 2 lists a number of Cu hyperaccumulators.

The accumulation of the other three metals by thrift showed no inordinately high concentrations of metals in the plant tissue. However, thrift is classified as an indicator plant for both Zn and Cu. This is reflected in the concentrations found in the plants growing on Soil A and to some extent on Soil B for Cu. This was not the case for Soil C, with plant concentrations being much less than those found in the soil.

Baker (1981) found concentrations of up to 7,500 mg/kg Zn and 50 mg/kg Cu in the plant tissue. These values far exceed (except for Cu) those found in

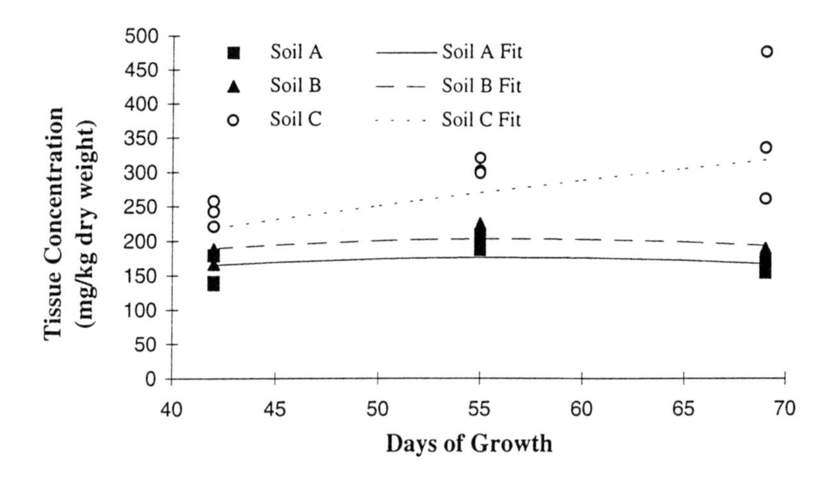

FIGURE 1. The uptake characteristic of Zn in thrift.

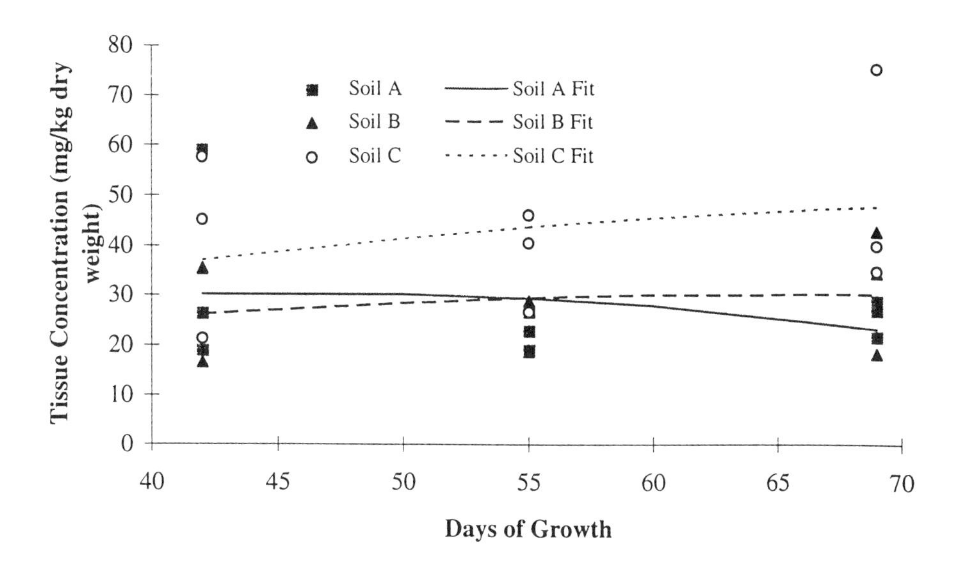

FIGURE 2. The uptake characteristic of Cu in thrift.

this trial; however, the soil concentrations were extremely high in this case, and the sources of lead were nonanthropogenic (Baker 1981). When the values from the current trial and those of Baker (1981) are converted to an accumulation factor (concentration of metal in plant/concentration of metal in soil), there are significant differences. The accumulation factors in the current trial for thrift were 0.93 (Zn), 0.94 (Cu), 0.09 (Pb), and 3.5 (Cd) for Soil A. For Soil B the accumulation factors were 0.79 (Zn), 0.67 (Cu), 0.09 (Pb), and 0.85 (Cd); and for Soil C the accumulation factors were 0.17 (Zn), 0.17 (Cu), 0.05 (Pb), and 0.43 (Cd). The

TABLE 2. Hyperaccumulators of copper.[a]

Plant Species	Maximum Tissue Concentrations (mg/kg)
Ipomoea alpina	12,300
Lindernia perennis	9,322
Haumaniastrum katangese	8,356
Bulbostylis mucronata	7,783
Pandiaka metallorum	6,260
Gutenbergia cupricola	5,095

(a) Concentrations in the aboveground plant parts; adapted from Baker and Brooks (1989).

accumulation factors for the data reported by Baker (1981) were 0.3 (Zn), 0.57 (Cu), 0.13 (Pb), and 0.08 (Cd). The differences in these values between the two trials, and within the current trial, indicate that soil type and source of metal strongly influence the uptake of metals. These differences between treatments in the current trial were significant at the 5% level, as determined from the statistical analysis.

The uptake of Zn by thrift in the current study reached its maximum concentration for Soils A and B at 50 to 55 days. However, for Soil C, the maximum concentration was not obtained during the time of the trial. From the data obtained in the current study, it is possible that the maximum concentration in the plant tissue in Soil C was reached at 70 days or longer (Figure 1). The maximum concentration of Cu in thrift was found to occur at 42 days for Soil A and 65 to 70 days for Soils B and C. The optimum growth period for Soil A may have been reached before 42 days; however, no harvests were conducted before this period.

It has previously been indicated that the accumulation of metals is greater when conducted in a glasshouse trial due to optimum conditions being maintained, as reviewed by Kelly (1994). Such conditions include differences in microclimate and soil moisture and the fact that the roots of container-grown plants grow solely in the contaminated soil. This was certainly the case in this trial. It was noted that, after 42 days, the root zone encompassed the whole pot. Because of this, uptake would have been enhanced.

It has been suggested that plants growing on metalliferous soil cannot prevent metal uptake but only restrict it. Hence such plants tend to accumulate metals in their tissues to varying degrees (Baker 1981). The subsequent uptake of metals varies according to the metal under consideration and the competition from other metals present. Based on this, the accumulator, indicator, and excluder mechanisms were conceived. Of the plants tested in the current study, only thrift (for Zn) was found to be an accumulator with an accumulation factor greater than 1. However, this was only with Soils A and B. On Soil C, the value was considerably less. This was a surprising result because thrift was reported as an excluder of Zn (Baker 1981) with a partition coefficient (i.e. ratio of metals in shoot to root) of less than 0.4. This literature value is, however, consistent with the current findings from Soil C. It is apparent, therefore, that the soil type and the total and available metal concentrations may play a role in determining whether or not a plant is an excluder or an accumulator. Although thrift showed an accumulation of Zn, none of the other plants showed signs of accumulation for any of the metals. This indicates that balsam, alyssum, and cabbage are excluders and will accumulate only after a certain threshold concentration is reached in the soil.

CONCLUSIONS

It is likely that the metal concentrations in the soil used in the current study may have been too low for hyperaccumulation to occur in thrift. Therefore, threshold concentrations of metal are likely to be necessary before particular plants are able to take up metals from contaminated soil with particular soil types.

TABLE 3. Plants that can hyperaccumulate more than one metal.

Plant Species	Metal Accumulated
Thlaspi alpestre	lead, zinc, nickel
Viola calaminara	nickel, zinc
T. rotundifolium	lead, nickel
T. caerulescens	nickel, zinc, cadmium
Alyssum montanum	copper, nickel
Haumaniastrum katangese	copper, cobalt

For meaningful comparisons to be made between studies, the soil type, source or origin of metal, and concentration, as well as species of metal, need to be reported. There is also a need for further applied research in the field, using soils with anthropogenic sources of metal contamination over a wide range of metal concentrations to assess the feasibility of this approach. Specifically, it would be valuable to ascertain whether single species (that can accumulate more than one metal) are effective for anthropogenic sources of metal such as those reviewed from the literature and summarized in Table 3. There are, however, further constraints to the approach of phytoremediation; and these include the ecology of hyperaccumulators. For instance, can they be grown in soil types and climates where there are anthropogenic sources of metal contamination and how many croppings of the plant would be needed to clean up a site to acceptable concentrations.

It is likely that plants may have a role in the reclamation of metal-contaminated soils, through both establishment of metal-tolerant plants for the rehabilitation of contaminated land and in the sequestering of large amounts of metals. There are potential gains for the mining and soil remediation industries to develop this technology for sites where the time required for remediation of metal contaminants is relatively long. Focused research is still necessary before the findings on hyperaccumulators can be translated to a field-scale remedial option for metal-contaminated soil.

REFERENCES

Baker, A.J.M. 1981. "Accumulators and Excluders: Strategies in the Response of Plants to Heavy Metals." *Journal of Plant Nutrition* 3(1-4): 643-654.

Baker, A.J.M., and R. R. Brooks. 1989. "Terrestrial Plants Which Hyperaccumulate Metallic Elements: A Review of Their Distribution, Ecology and Phytochemistry." *Biorecovery* 1: 81-126.

Brooks, R. R. 1994. *Biological Methods of Prospecting for Minerals*. John Wiley and Sons, New York, NY.

Kelly, R. J. 1994. "The Uptake and Accumulation of Heavy Metals by Selected Plant Species: Implications for Soil Remediation." BSc. (Honors) Thesis. The University of Sydney, Sydney, Australia.

Kelly, R. J., and T. F. Guerin. 1994. "The Potential of Using Plants as an Innovative Approach to Soil Decontamination." *2nd National and Hazardous Waste Convention*, pp. 591-598. Australian Waste Water Association, Sydney, Australia.

McGrath, S. P., C. M. Sidoli, A.J.M. Baker, and R. D. Reeves. 1994. "Using Plants to Clean-up Heavy Metals in Soils." *15th World Congress of Soil Science*, pp. 310-312. Acapulco, Mexico, July 10-16, Vol. 4a.

Morrison, R. S., R. R. Brooks, R. D. Reeves, and F. Malaisse. 1979. "Copper and Cobalt Uptake by Metallophytes From Zaire." *Plant and Soil 53*: 535-539.

Remediation of Contaminated Soils and Sludges by Green Plants

Scott D. Cunningham, William R. Berti, and Jianwei W. Huang

ABSTRACT

The potential of green plants to remove, contain, or render harmless contaminants in soils and sludges is actively being explored in an increasing number of laboratories throughout the world. This approach, which has been termed "phytoremediation," exploits plants, soil amendments, and plant-associated microbiota to remediate contaminated soils. As an in situ stabilization technique, soil amendment with fertilizers, biosolids, or certain industrial by-products alters the chemical and physical nature of the contaminant in the soil matrix, thus reducing its availability to biological processes. The site is then vegetated with plants that can (1) grow in the resulting soil matrix; (2) reduce leaching through the soil profile by absorbing, sequestering, or degrading residual contaminants in the soil solution; and (3) minimize wind and rain erosion. The process is known as "phytostabilization," or simply "site stabilization," and borrows heavily on mine reclamation techniques. As a site decontamination technique, the soil is treated to increase the availability of the contaminant to biological processes and then planted with plants that (1) accumulate the contaminant and are harvested for further pollutant destruction, sequestration, or reclamation or (2) use plant or plant-associated microbial processes to destroy the pollutant in situ.

INTRODUCTION

With increasing clarity, we have begun to appreciate that plants play a remarkable role in maintaining the quality of our lives. In addition to providing their traditional functions of food, shelter, heat, clothing, and medicine, plant life also functions in the micro, macro, and global processes of elemental cycling, which is critical to the quality of our water, air, and soil. Plants accumulate and concentrate certain inorganic compounds. Plants have also evolved to both produce and metabolize a wide range of natural toxins in their struggle to survive in a hostile environment. Plants support microbial communities both external to the root surface, as well as internal to the roots, stems, and leaves. These microbial associations allow for even greater degradative metabolic capacities.

As our appreciation of these processes has increased, researchers have begun to view plants with an eye toward the development of "minimal disturbance" remediation systems. Developing and exploiting sophisticated agronomic principles currently allows us to take soils affected by sea salt, acidity, alkalinity, ore outcroppings, erosion, and water logging and convert them to agronomically productive land. Exploiting these capabilities to remediate industrially contaminated sites is a logical extension.

Plant-based remediation systems for water-borne contaminants are well advanced. Principles learned from wastewater handling and overland contaminated water systems have been extended and adapted to include a variety of plant-based systems in which the plants float (e.g., duckweed and water hyacinth) as well as systems in which the plants are rooted in soil or sediments (e.g., constructed wetlands and reed beds). The use of plants to remove airborne "contaminants" is also well known. Lavoisier's classic bell-jar experiments alternating a mouse and plant in the same environment clearly demonstrated the interdependence of mammalian respiration and plant growth. Plants have large leaf surface areas, waxy coatings, and sophisticated gas-exchange mechanisms. Plants and their root-associated microbial flora are being exploited to absorb and metabolize a wide range of airborne pollutants. The first field-based subterranean biofilters have been in operation for more than 40 years in Europe, and the nature and rate of absorption of pollutants into leaves is increasingly the subject of research by those involved with urban air quality.

Despite the increasingly sophisticated use of plants in the remediation of contaminated waters and air, the use of plant-based systems to remediate contaminated soils and sludges is only just beginning. All soil and sludge remediation processes accomplish one of two objectives. They reduce the hazard presented at a site by either (1) sequestering the pollutant into the contaminated matrix or (2) removing or destroying the pollutant in the matrix, leaving behind a decontaminated material. Engineering examples of the first technique include all technologies roughly grouped as "isolation" or "containment" techniques including dig and haul, capping, vitrification, and stabilization. Engineering techniques that fall into the second category include many thermal techniques, soil washing, bioremediation, and vacuum extraction.

Like these engineering technologies, plant-based remediation techniques can be employed to accomplish either containment/isolation or pollutant removal. The overall term that is applied to both techniques is "phytoremediation," where "phyto" is derived from the Greek prefix meaning plant. This paper will use "phytoextraction," "phytostabilization," and the generic "phytoremediation." "Phyto" is also used with many standard engineering and scientific terms to describe potential remediation techniques such as phytoabsorption, phytosequestration, phytolignification, and phytodegradation. Other terms similarly employed include "green remediation," "botano-remediation," and, in one variation, "biomining."

The following is a brief listing of some important biological and chemical processes that occur in soils and would be expected to be important in site remediation. This list is grouped into processes that lead to increased stabilization and processes that lead to site decontamination.

Stabilization:

Plant processes that aid in stabilization:

1. transport of ions across root-cell membranes
2. water flux to the plant driven by plant transpiration
3. absorption of organics into the roots
4. entrapment of organic in the lignin fraction of plants (lignification).

Soil processes that aid in stabilization:

1. biochemical fixation (humification) — enzymatic incorporation into humus
2. chemical fixation: precipitation
3. physical fixation: solid-state diffusion into soil structures (clays, organic matter, etc.), formation of oxide coatings.

Decontamination (pollutant destruction and/or extraction):

Plant processes that aid in pollutant destruction and/or extraction:

1. bioactive microbial biofilm around plant roots (rhizosphere)
2. plant and microbial-produced surface-active agents and chelates
3. fungal symbionts on roots that extend out into the soil and increase soil-to-surface area ratios and provide additional enzymatic capacity
4. root, stem, and leaf enzymatic metabolic activities for detoxification
5. uptake of cations and some anions into the root
6. translocation of absorbed ions from roots to shoots
7. solar-driven solution flux from soil, through roots into plant shoots
8. partitioning of lipophilic organic molecules into roots.

Soil processes that aid in pollutant destruction and/or extraction:

1. agronomic practices that provide air, nutrients, surface area disruptions, crop residue cycling, chemical fluxes, and microbial stimulation
2. bulk soil microbial degradation
3. bulk soil chemical degradation (e.g., on catalytically active clay surfaces)
4. wetting and drying cycles (reduction and oxidation)
5. chemical/biochemical — general hydrolytic, substitution, and elimination reactions.

As the two preceding paragraphs suggest, phytoremediation is applicable to a wide variety of contaminated matrices. If plants can be grown and maintained, then phytoremediation may have potential. As with all remediation technologies, however, there are inherent limitations to the technology. The first such limitation is that rooting depths are not infinite. Roots usually go as deeply as they need to find water and nutrients. Depths are generally limited to a range of 1 to 10 m depending on pollutant, stratigraphy, crop, and permeability of the matrix to roots. There is a considerable difference in rooting morphologies among plants that can be exploited; however, rooting densities often are dramatically skewed in a soil profile. Most root mass occupies the top 30 cm, which may not coincide with pollutant distribution.

The second limitation is that roots are living and have significant environmental limitations for pH, texture, temperature, osmotic pressure, moisture, and oxygen. Obtaining appropriate growth parameters may be relatively easy on many contaminated sites with perhaps the one following exception: roots require oxygen for respiration, hence they do not tend to go deeply into anaerobic zones. Many sites contaminated with biodegradable organics tend to become anaerobic due to microbial activity. All rooting systems, even those that are well adapted to anaerobic zone interfaces, do not penetrate deeply into such zones. Plants with sophisticated adaptations to help translocate oxygen from the atmosphere to their roots find that they can meet their water, nutrient, and structural support requirements by only barely penetrating the anaerobic zone. Rice, salt marsh grasses, and even mangrove roots penetrate less than half a meter into such anaerobic environments. One might expect that trees might require deep rooting systems as anchorage against wind storms; however, in these situations, trees tend to develop alternative solutions to the anaerobic dilemma. The knees of swamp cypress and the aerial roots of many others are two such adaptations visible in bogs and swamps. The biology and botany suggest that tackling even relatively shallow anaerobic zones may require either other technologies or significant engineering input.

Another inherent limitation of using plant-based systems is that plants provide an imperfect barrier against leaching. Rooting systems and plant transpiration can dramatically reduce contaminant flux through a soil profile by dewatering the soil profile in addition to contaminant uptake; however, without a secondary containment system, phytoremediation should be targeted for only those pollutants and sites where the potential for groundwater contamination is low. There are exceptions for certain climatic, hydrologic, and soil conditions, but in general highly mobile inorganic salts as well as water-soluble organics are not considered appropriate candidates for phytoremediation without an engineered secondary containment system.

A general limitation of all biological systems in site cleanup is that for the plant or plant-associated microflora to interact and remediate the contaminant, the contaminant must be biologically available. This constraint suggests that many of the same endpoint limitations incumbent in microbial forms of biological remediation will also apply in plant remediation systems. Pollutants sequestered into clay lattices, absorbed into soil humic fractions, and occluded by oxide coatings are not accessible to most biological processes and are extremely stable. Although they may be detected by exhaustive solvent extraction (organics) or sample acid digestion (inorganics), the hazard these materials pose to the environment or human health is unknown and under debate. In some cases, a particular pollutant-matrix can be proven to have only limited biological availability to microbial, plant, arthropod, amphibian, and mammalian system, yet still be deemed "unacceptably contaminated" by chemical analytical techniques. Bioavailability relative to both inherent hazard and the development of biological remediation technologies is currently an active research topic in remediation and the subject of task forces focused on the determination of "environmentally acceptable endpoints." Two such soil contaminants under active discussion involve petroleum hydrocarbon residues and lead (Pb) in soils.

The last general constraint is that all forms of phytoremediation of soils and sludges that involve decontamination may require more time than many engineering solutions (especially involving excavation and removal). Situations that pose an imminent risk to human health or further environmental upset should be considered less amenable to this form of remediation.

Despite these limitations, we, along with many others in the research, engineering, and remediation communities, believe that plant-based systems can provide low-cost, low-impact, visually benign, and environmentally sound strategies that will continue to develop into a technology applicable for many contaminated soils and sludges.

INORGANIC POLLUTANTS

Inorganic Chemistry in Soil

Most inorganic pollutants have multiple chemical and physical forms in the soil environment. Most readers may be aware that not all forms are equally hazardous, nor are all forms equally amenable to uptake into plants. It is well recognized by both the engineering and regulatory communities that certain elements pose differential risks based on their oxidation state and/or their degree of alkylation (e.g., Cr, As, Se, and Hg). The principle can further be extended to most elemental contaminants and is more subtle than is commonly recognized. The chemistry of the contaminant and its interaction with the soil/sludge matrix determine its inherent hazard, the potential success of phytostabilization relative to phytoextraction, as well as the success of a wide range of engineering-based remediation techniques. In the following text, the authors will outline both phytostabilization and phytoextraction of inorganics. We will address in detail one element, Pb, although much of the information presented pertains to other elements as well.

Over the last 3 years, we have explored many methods for characterizing Pb-contaminated soils. We have examined approximately 30 Pb-contaminated soils attempting to chemically, physically, and biologically predict the hazard that a particular soil would present to environmental and biological systems. One of the most useful findings of this study has been that a particular sequential chemical extraction yields a good approximation of "hazard" as measured by the Toxicity Characteristic Leaching Procedure (TCLP), plant uptake, and mammalian in vitro bioassays. Sequential chemical extraction techniques involve exposing a sample of the contaminated soil to a series of solutions of increasing chemical harshness. After each step, the Pb in the extracting solution is measured. The last step consists of a complete nitric-perchloric acid sample digestion.

The Pb concentration in each fraction is calculated as a percent of the total Pb in the sample.

We performed sequential chemical extractions on 1-g samples, using 40 g of each of nine solutions and conditions in the sequence outlined in Table 1.

The results of sequential chemical extractions have proven useful to us in three ways: (1) as a measure of relative hazard posed by the Pb in the matrix; (2) as a predictor of the success of phytostabilization relative to phytoextraction; and (3) as a first cut to suggest the applicability of engineering technologies that might be applied at the site including soil washing, electroosmosis, and stabilization. These points are perhaps best explained relative to Figure 1.

In this figure, results from the 9-step sequential chemical extraction technique are depicted under two chemically extreme conditions. From both the figure and an intuitive chemical sense, one would expect the Pb nitrate in the sand to readily leach, be available for uptake into plants, and be available in a mammalian digestive tract (or 100% bioaccessible). In contrast, Pb metal in the clay sample would not be expected to leach, nor be available to plants or mammalian stomachs as it is available only in the solution phase under decidedly nonbiological conditions (nitric-perchloric acid digestion). Regulations do not currently account for these dramatic differences in chemical forms of Pb. Regulations based on assays of total Pb in the sample significantly underestimate the toxicity of the

TABLE 1. A 9-step sequential extraction protocol used to differentiate the chemical forms of Pb in a soil or sediment sample. (Step 4 is omitted for samples with a soil pH less than 7.) Contaminants that are removed in the early fractions are relatively mobile, subject to loss by leaching, plant uptake, and mammalian bioavailability upon ingestion. Contaminants in the latter fractions are relatively environmentally inert.

Fraction Number	Fraction	Procedure
1	Water-soluble	Deionized water and mix for 16 h
2	Exchangeable	0.5M $Ca(NO_3)_2$ at soil pH for 16 h
3	Specifically adsorbed	0.05M $AgNO_3$ + 0.1M $Ca(NO_3)_2$ for 16 h
4	Carbonates	1M $NaCH_3COO$ at pH 5 for 5 h
5	Mn oxides	0.1M $NH_2OH{\cdot}HCl$ + 0.1M HNO_3 for 30 min
6	Organic	0.1M $Na_4P_2O_7$ at pH 10 for 24 h
7	Noncrystalline Fe oxides	0.25M $NH_2OH{\cdot}HCl$ + 0.25M HCl for 30 min
8	Crystalline Fe oxides	0.4M $NH_2OH{\cdot}HCl$ in 25% v/v CH_3COOH mixed periodically in a 90°C water-bath for 6 h
9	Residual	Concentrated $HNO_3/HClO_4$ microwave digestion for total metal analysis

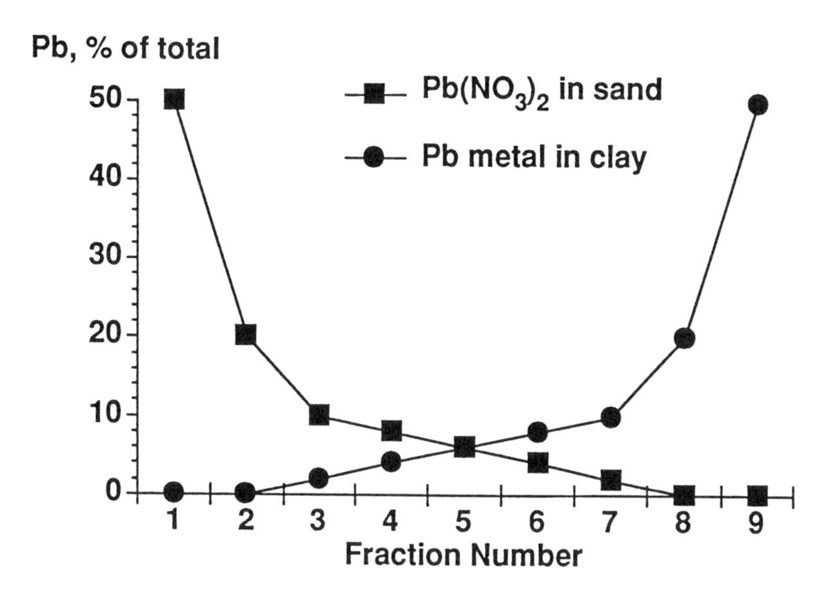

FIGURE 1. Idealized sequential chemical extraction of lead nitrate added to sand and lead metal added to clay. See Table 1 for the extraction protocol used to generate each fraction.

Pb nitrate in sand, and overestimate the toxicity of Pb metal in clay. It is also apparent from Figure 1 that the contaminated sand is a candidate for either soil washing or phytoextraction and the contaminated clay is a candidate for phytostabilization and will not be amenable to many solution-phase Pb extraction techniques.

Figure 2 shows sequential chemical extraction of three different industrially contaminated soils. The results suggest that, like the examples in Figure 1, there is a wide variety of inherent difference in soil Pb in real soils. These three soils pose very different opportunities and hazards. About 50% of the 2,500 mg Pb kg^{-1} in Suscon CDA soil and 15% of the 500 mg Pb kg^{-1} in the Weyerhauser soil occur in the exchangeable and specifically adsorbed fractions. These fractions are easily leachable and susceptible to phytoextraction and soil washing techniques. The remainder, however, will be more difficult to remove. Almost 60% of the approximately 1,500 mg Pb kg^{-1} in the C.W. SWMU-57 soil is in the carbonate fraction that is available under slightly acidic conditions. This soil would be predicted to have Pb forms unavailable to plants without acidification, but would fail TCLP and have Pb bioavailable to children upon soil ingestion. All three of these soils are candidates for phytostabilization.

Phytostabilization of Pb-Contaminated Soil

Under a phytostabilization strategy, the mobility of inorganic contaminants is reduced by the addition of soil amendments that reduce contaminant solubility.

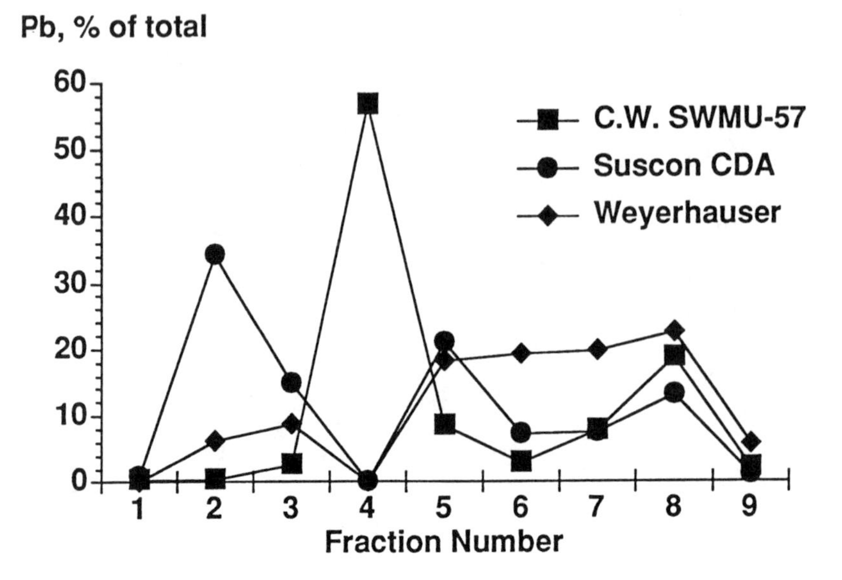

FIGURE 2. Results of sequential chemical extraction of three soils contaminated with lead. See Table 1 for the extraction protocol used to generate each fraction.

We have used a variety of alkalizing agents, phosphates, mineral oxides, organic matter, and biosolids to render Pb more insoluble and unavailable to leaching, mammalian ingestion, and plant uptake. We choose amendments to transform the highly soluble and potentially hazardous forms of Pb, which occur in fractions 1 to 3 of the sequential chemical extraction profile of Figure 2, into less soluble and less hazardous Pb forms. These amendments should be low cost and nonhazardous and should help condition the soil for better plant growth.

The success of our amendment efforts is shown initially by a shift of Pb from early fractions (1-4) into later fractions (5-9). This chemical alteration is then followed by quickly establishing a plant cover and maximizing plant growth. The amendments sequester the pollutant into the soil matrix and the plants keep the stabilized matrix in place, minimizing wind and water erosion. Additionally, the roots themselves absorb any residual Pb in the soil solution, reducing its susceptibility to leaching and bioavailability to other organisms. Growing plants also reduce the potential for groundwater contamination as the total flux of water flowing downward through the soil is reduced as a result of root uptake and plant transpiration of water into the atmosphere.

Figure 3 represents changes that occur in the C.W. SWMU-57 soil when amended with Fe oxyhydroxides (e.g., Iron Rich 101, which is a DuPont mineral by-product resulting from manufacturing TiO_2) or phosphate fertilizer. The nine-step sequential chemical extraction clearly shows significant reduction in presumptive Pb hazard by addition of these materials. This change in soil-Pb

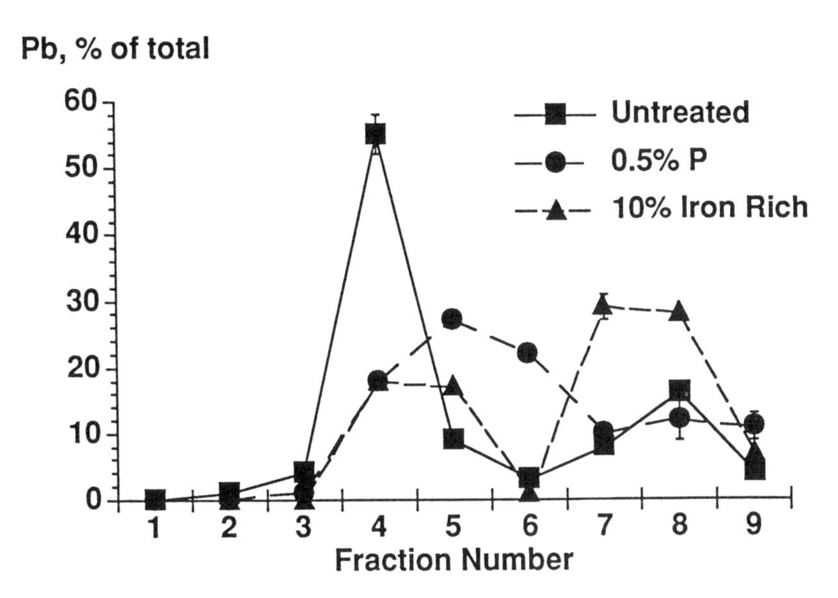

FIGURE 3. Results of sequential chemical extractions of C.W. SWMU-57 soil: untreated, treated with 0.5% P (as KH_2PO_4), and treated with 10% Iron Rich 101. (Percentages are weight/weight and based on dry weight.) See Table 1 for the extraction protocol used to generate each fraction. Data from Berti and Cunningham (1995).

chemistry is a result of the fixation of available Pb forms by both P and Fe oxides. As a consequence of the amendments, these soils have a significantly lower Pb hazard as reflected by TCLP, plant bioavailability testing, and an in vitro mammalian bioaccessibility test. The TCLP results are listed in Table 2. Plant availability tests were conducted under greenhouse conditions with red-top (*Agrostis alba*, a wide-ranging conservation grass for the northeastern United States). Data from unpublished studies show that soil amendments that produce significant shifts in Pb sequential extraction can also reduce shoot Pb concentrations by more than 90%. In addition, these soil amendments increased the tilth of the soil and were generally beneficial to plant growth and yield. Similar reductions were seen in presumptive mammalian Pb bioavailability upon addition of these amendments. Lead bioaccessibility, measured using an in vitro simulated mammalian digestive tract assay from a method by Ruby et al. (1993), was reduced by more than 45 and 90%, respectively, in soil amended with phosphate fertilizer and Iron Rich compared to untreated soil.

Although a relatively new concept, phytostabilization borrows from much previous research in the reclamation and revegetation of drastically disturbed lands and land disposal of sludges and biosolids. A number of metal-tolerant plants have been used to vegetate and control soil erosion on metal mine tailing and waste piles (Bradshaw & Chadwick 1980). One of the benefits of these plants is that

TABLE 2. TCLP Pb and TCLP pH determined in C.W. SWMU-57 soil amended with P fertilizer (added as KH_2PO_4) Iron Rich 101, organic carbon, and biosolids. All amendments were added on a weight/weight, dry weight basis (Berti & Cunningham 1995).

Treatment	TCLP Pb (mg L^{-1})	TCLP pH
Untreated	28.8 ± 3.3	5.05 ± 0.02
0.005% P	21.8 ± 1.9	5.29 ± 0.26
0.05% P	5.0 ± 1.9	5.14 ± 0.04
0.5% P	0.1 ± 0.1	5.18 ± 0.03
1% Iron Rich 101	18.0 ± 2.0	5.12 ± 0.04
10% Iron Rich 101	0.7 ± 0.5	5.29 ± 0.05
2% Sphagnum Peat Moss	15.4 ± 5.9	5.12 ± 0.08
2% Natural Humus	17.1 ± 2.7	5.08 ± 0.03
2% Seaford Works Biosolids	8.0 ± 0.7	5.04 ± 0.02

most of the metal they take up remains in the roots of these plants, consequently these areas can be used in grazing management strategies. Generally, animal grazing on these restored mine sites is limited to short time periods, and animals are rotated to uncontaminated land for much of their feeding requirements.

Metal-contaminated sludges can pose a significant problem in upland disposal sites. Brandon et al. (1991) describe soil amendments of lime, coarse limestone gravel, and horse manure for acidified, metal-contaminated dredged material. These sites are then planted with metal-tolerant species. Untreated control plots are barren even after 6 years and produce an acid metal-rich surface runoff of pH 3.5 that exceeded water quality standards. Plots receiving soil amendments and acid/metal-tolerant plants produced an acceptable surface runoff.

Phytoextraction

Phytostabilization is a technology that is moving rapidly from the research to the development phase, with field tests in progress. Phytoextraction is more ambitious and technically more difficult; as such it is primarily still a research topic, albeit a dozen field tests are in progress throughout the world. The processes involved in phytoextraction are shown in Figure 4. For phytoextraction to be an economically viable technology, the pollutant must be available to the plant root and plant roots must take it up. We believe that translocation from root to shoot must occur for both ease of harvesting and to minimize worker exposure during harvest, although some suggest root harvesting techniques could be feasible under certain conditions. After harvesting, a biomass processing step occurs to recover the metal and/or further concentrate the metal by thermal, microbial, physical, or chemical means to further decrease handling, processing, or landfilling costs.

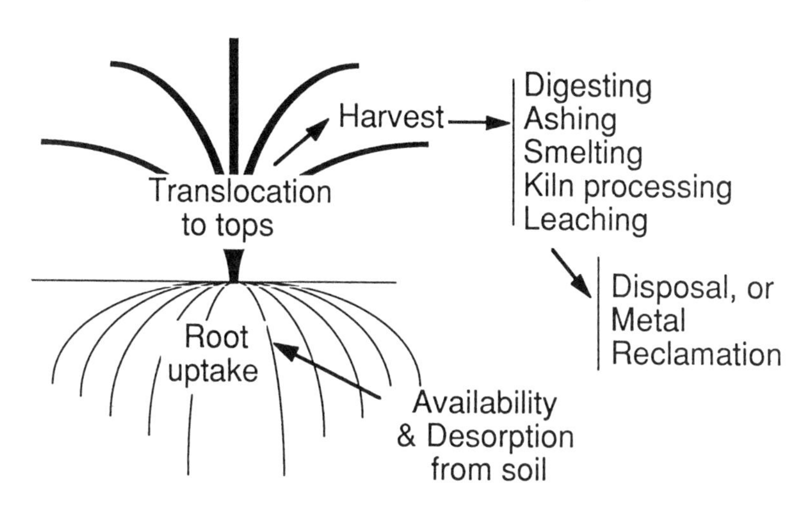

FIGURE 4. Phytoextraction schematic (adapted from Cunningham and Berti 1993).

Each inorganic pollutant presents its own unique set of problems and opportunities. Elements range in toxicity, chemistry, physical behavior, economic value, and essential biological requirement status. Phytoremediation is actively being considered for perhaps a dozen elements throughout the world. The race to reduce the technology to practice seems to depend on (1) inherent biology (all plants have significant uptake mechanisms for elements that are essential to plant health); (2) regulatory status (higher regulatory acceptable soil limits and longer time frames); (3) chemistry (some elemental forms are extremely soft Lewis acids, which means they are chemically "stickier" to soil surfaces and inside plants); and (4) additional economic incentive (the value of the reclaimed metal might be significant). Combining these factors, it is our opinion that phytoextraction will first be demonstrated on a field scale for Ni, Zn, and Cu. Other trace elements including Se, Cd, Pb, and radionuclides will follow. Field tests for certain elements have begun (Baker et al. 1991), but there remain significant technical hurdles before full-scale implementation is widespread.

Plants growing in common garden soil acquire perhaps two dozen elements in their shoots. Plant tissue concentrations of these elements range widely from trace levels to as high as 5% by dry weight (a variation of over seven orders of magnitude). For the most part, plants take up large amounts of elements that are required for growth, and only small amounts of toxic elements that could harm them. Some nonessential elements appear in large amounts that are not particularly harmful (e.g., Si and Na), but many of our target pollutants are only 0.1 to 100 mg kg^{-1} in a plant (Jeffery 1987). The final goal of phytoextraction techniques is to remove contaminated metals from the soil. Metal removal rates are dependent on both the quantity of plant biomass removed as well as the metal concentration in the plants. Total metal removal is calculated by multiplying the weight of biomass per area times metal content in the biomass times the number of harvests. An engineering cost evaluation determined that, for site decontamination to occur within 10 harvests, plant parameters must include both a high yield of biomass and metal accumulation greater than 1 to 3% metal by dry weight. Most researchers in this area are attempting to find, breed, or engineer plants with metal concentrations in the range of 1 to 3%.

Conceiving and producing plants that maintain the extreme levels of metal uptake, translocation, and tolerance that are toxic to biological processes might be thought impossible were it not for the existence of a small group of remarkable plants called hyperaccumulators (Brooks et al. 1977). Although uncommon, hyperaccumulating plants are widespread throughout the plant kingdom (Baker & Brooks 1989) and occur across multiple families and genera. A small tree, *Sebertia acuminata,* occurs naturally on a nickel ore outcropping in New Caledonia. The tree has sap that is 25% by dry weight nickel (Jaffre et al. 1976). One ecotype of the *Thlaspi* species (a member of the Brassica family) grows in soils high in Zn. This plant accumulates 3% of its total dry weight in Zn (Reeves & Brooks 1983a). A list of some of these remarkable plants with their metal accumulation capabilities is in Table 3.

Lead, which we are interested in phytoextracting, appears to be a poor metal to target for this technology. Lead is not a required element for either plants or animals. Because it forms strong bonds with soil minerals and organic matter, it is difficult for plants to extract it from soils and into their roots. It easily complexes with plant nutrients once inside the roots, reducing the amount translocated from roots to shoots.

As a consequence, it is surprising to find that shoots of *Thlaspi rotundifolium* growing in soil contaminated with Pb has been reported to contain as high as 8,200 mg Pb kg^{-1} (Reeves & Brooks 1983b). Due to its low growth rate and small

TABLE 3. Metal concentrations (on a dry weight basis) in plant tops of known metal hyperaccumulators.

Metal	Plant Species	Concentration (mg kg^{-1})	Reference
Cd	*Thlaspi caerulescens*	1,800	Brown et al. 1994
Cu	*Ipomoea alpina*	12,300	Malaisse et al. 1979
Co	*Haumaniastrum robertii*	10,200	Brooks et al. 1977
Pb	*Thlaspi rotundifolium*	8,200	Reeves & Brooks 1983b
Mn	*Macadamia neurophylla*	51,800	Jaffre 1979
Ni	*Psychotria douarrei*	47,500	Jaffre 1980
Zn	*Thlaspi caerulescens*	51,600	Brown et al. 1994

biomass, however, this species is not well suited for Pb phytoextraction. In 1991 we instituted a field and lab program to find or develop plants that absorb, translocate, and tolerate 1 to 2% of Pb on a dry weight basis. As we discovered, these are ambitious goals because, even on heavily contaminated soils, most vegetation had less than 50 ppm Pb in the shoot tissue. Our efforts have three phases: (1) find an existing plant by examining the flora of old mine sites and our older Pb-contaminated industrial sites; (2) describe the plant physiology of Pb uptake, translocation, and tolerance and eliminate the rate-limiting step by the use of engineering, chemical, or physical adjuncts; and (3) create a better plant through selection, breeding, or molecular biological techniques. We have collected perhaps half a dozen plant species, including common ragweed (*Ambrosia artemisiifolia*), Asiatic dayflower (*Commelina communis*), and Indian hemp or hemp dogbane (*Apocynum cannabinum*) with a range of 500 to 1,000 ppm Pb in their shoots. The Pb is translocated poorly, however, sequestered mostly in the lower stem of these plants.

As part of the physiological characterization, we are conducting screening trials in hydroponic solutions at a 4-ppm ambient Pb concentration. We have found that experiments with seedlings grown in this solution for 2 weeks accurately reflect relative Pb uptake, translocation, and tolerance in Pb-contaminated soil. The results presented in Table 4 show that there is roughly a 10-fold variation in uptake rates and 50-fold difference in translocation rates (root to shoot) between species. The ideal plant would take up Pb like a sunflower root (accumulating nearly 2% of the root dry weight when grown in only a 4 ppm solution), and translocate it at rates exceeding that of corn, which has the highest shoot concentration (0.2% Pb) that we have found to date. We have increased uptake and translocation of Pb through the use of chelate solutions and electroosmosis (technology that uses electric current to translocate metals in soils and, potentially, plants), but ideally we would like to have a plant capable of more robust metal uptake and translocation than we have found so far. Many hyperaccumulators, in addition to having yields of 1 to 3% by dry weight, have superior translocation mechanisms from root to shoot.

Although we continue to search for a Pb-hyperaccumulating native plant with high biomass production, molecular techniques to engineer our ideal plant appear necessary. Hyperaccumulators have the metal-accumulating characteristics that we seek, but lack the biomass production, adaptations to current agronomic techniques, and physiological adaptations to the climatic conditions at many contaminated sites in industrial areas. We, as well as other researchers, are looking toward altering the genetic makeup of higher biomass plants to produce the ideal plant for phytoextraction. The use of "genetic engineering" is not without precedent in metal accumulation. Using mutation techniques, scientists have succeeded in identifying various metal-accumulator mutants. For examples, a single-gene pea mutant (*Pisum sativum*) has been identified from chemically mutagenized parent pea seeds. Depending on Fe levels in root growth medium, this pea mutant can accumulate 10- to 100-fold higher Fe than its parent pea plants (Welch & LaRue 1990; Grusak 1994). Recently, a Mn-accumulating mutant has been identified by screening chemically mutagenized *Arabidoposis* plants

TABLE 4. Lead concentrations (on a dry weight basis) in shoots and roots of selected plant species grown in nutrient solutions[a] containing 20 μM Pb as $Pb(NO_3)$.

Plant Species	Pb Concentration ($mg\ kg^{-1}$) Shoots	Roots	Shoot/ Root Ratio
Dicots			
Sunflower	85	19,900	0.004
Goldenrod	96	8,130	0.012
Ragweed	110	5,073	0.022
Lupin	189	7,830	0.024
Monocots			
Sorghum	150	3,730	0.040
Wheat	180	5,940	0.030
Bermuda Grass	420	9,340	0.045
Corn	490	2,110	0.230

(a) Nutrient solution composition was the same as that of Huang et al. (1993), except that P concentration was 10 μM. The nutrient solution was continuously pumped using a microprocessor-controlled multichannel cartridge pump (Cole-Parmer) into an 8-L plastic container that held four plants. All plants were grown in nutrient solution for a week before the Pb exposure and were harvested after 2 weeks of Pb treatment (Huang & Cunningham, unpublished results).

(Delhaize et al. 1994). This Mn mutant accumulated 10-fold greater Mn in the leaves than did the wild types. These results indicate mutation techniques are useful tools for generation of metal-accumulating phenotypes. This technique may be useful to create other higher biomass metal hyperaccumulators.

The success of phytoextraction depends on the use of an integrated approach to soil and plant management. The disciplines of soil chemistry, soil fertility, agronomy, plant physiology, and plant genetic engineering are currently being used to increase both the rate and the efficiency of metal phytoextraction. The technology is still in a developing stage, but it holds a promising potential.

ORGANIC POLLUTANTS

Introduction to Organic Phytoremediation

To a large degree, those involved with plant-based remediation of xenobiotics (compounds foreign to biological systems) use a different terminology from that used in the parallel inorganic community. The terms most frequently encountered in organic remediation are the generic term "phytoremediation" as well as "rhizodegradation" or "degradation in the rhizosphere." There are two reasons for this terminology shift. The terms "phytoextraction" and "phytostabilization" imply a knowledge of the fate of the contaminant that is more difficult to obtain in organic uptake/degradation studies. The use of the generic term "phytoremediation" reflects the imprecision of both the analytical techniques and our understanding of the complex organic fate pathway in the environment. The term "rhizodegradation" reflects the belief that plants are generally thought of as being merely "constructive" and having little metabolic capacity, whereas the microbial community that flourishes in the root zone of actively growing plants is more "destructive," offering significantly increased metabolic potential.

The regulatory community traditionally has demanded a relatively clear understanding of the fate of an environmental pollutant prior to acceptance of a nonhazardous status. Extensive literature from the fate of pesticides in the environment shows that even rough mass balances are difficult to obtain due to multiple physical and biological processes occurring in the soil environment. Unlike inorganic pollutants where total sample digestion procedures provide excellent mass balance and fate information, organic contaminants are often in plant and soil matrices high in background organic materials. A general consensus of the regulatory community suggests that "disappearance" may be becoming acceptable. Initial acceptance of a process generically known as "natural attenuation," "natural bioremediation," or "bioattenuation" for some organic pollutants is developing. Some mass balance information is obtainable in soil systems through the use of sophisticated enclosed microcosms and radiolabeled compounds. This work is being done with specific priority pollutants; however, with complex mixtures of chemicals (e.g., PCBs and petroleum distillates and their by-products), this will prove impossible. The acceptance of these inexact disappearance mechanisms, the reliance on biological hazard assays, and the development and coordination of governmental policies will hasten the development of the phytoremediation of organic compounds.

As a general rule, regulatory protocols demand that soil characterization be determined by an analytical technique based on an exhaustive solvent extraction. This technique was codified into law at a particular moment in the evolution of analytical chemistry and represents a historical, and in some ways unfortunate, artifact. Many contaminated sites are the result of historical events that are decades old. There is a significant body of research on radiolabeled pesticides that suggests that as organic compounds in soils age, their association with the soil matrix changes physically, chemically, and biologically. With soils that have equilibrated with the organic material, exhaustive solvent extraction techniques generally underestimate the total amount of material present in soil samples (Alexander 1994) but overestimate the biologically available, or hazardous portion. Multiple experiments on soils up to 5 years old have shown that exhaustive solvent extraction protocols extract increasingly less of the total compound present as the soil ages (Pignatello 1989). Techniques that are considered more efficient at extraction of these "aged" contaminants are supercritical fluid extraction or high-temperature thermal desorption.

Figure 5 presents a generalized picture of a relatively recalcitrant pesticide in soil over time. The curves show that (1) the understanding of what occurs in soil is dependent on the assay technique; (2) the use of the various techniques produces increasingly divergent hazard assessments over time; and (3) with well-aged samples, solvent extraction represents neither a good indication of biological hazard (as indicated in this case by plant root injury) nor a total concentration of contaminant present in the soil.

One additional comment on analytical competency is appropriate before launching into a discussion of the phytoremediation of organic contaminants. The plant and soil science communities have been actively researching the fate of organics in the environment for 40 years. The regulatory requirements under the Federal Insecticide Fungicide and Rodenticide Act (FIFRA) have produced a wealth of information and a variety of sophisticated techniques to determine the fate of xenobiotics (compounds foreign to biological systems) in the environment. Xenobiotics have been applied to soil systems as both agronomic pesticides and

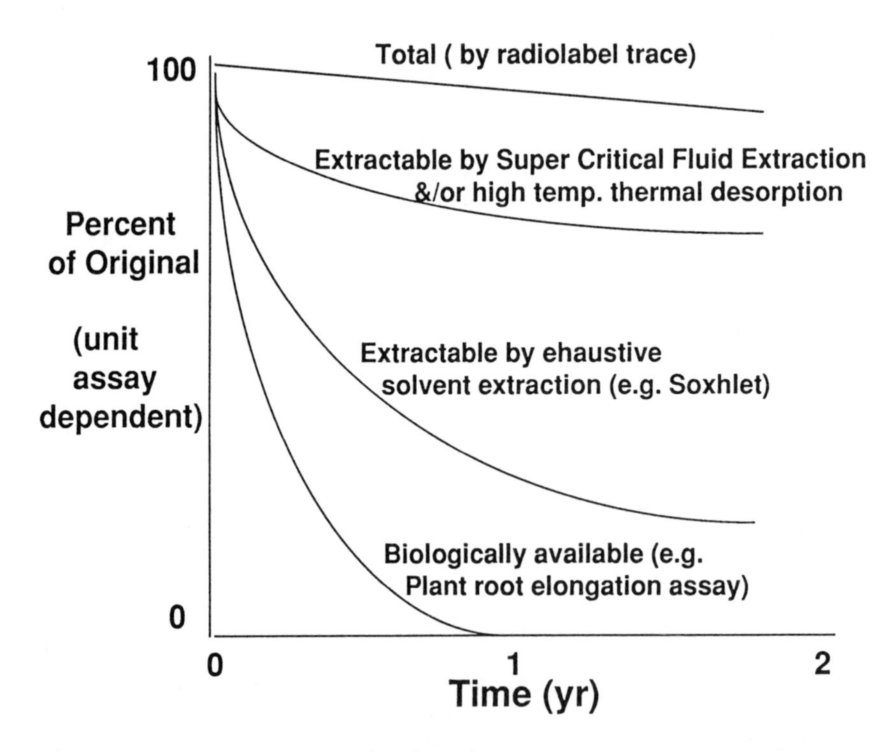

FIGURE 5. Time-dependent "disappearance" of a recalcitrant contaminant over time. Exhaustive solvent extraction techniques represent no good estimate either of compound remaining or of its biological availability/ hazard.

as a result of industrial accidents. Both situations are regulated by the United States Environmental Protection Agency (U.S. EPA), albeit different offices. Although the technical basis of many of the generic principles applied in the phytoremediation of organic contaminants is derived from pesticide research, the remediation of contaminated sites is expected to follow the same fundamental principles.

Target Pollutants

Not all organic compounds are equally accessible to plant roots in the soil environment. Roots have inherent ability to absorb organic pollutants based on the relative lipophilicity of the compound (Bell 1992). This property of the pollutant can be measured by its relative partitioning between an octanol phase and a water phase. It is often expressed as the log of the partitioning coefficient or log K_{ow}. Compounds with log K_{ow} values above zero partition preferentially from a water phase into roots (see Bell 1992). There are some differences between roots, but generally the higher a compound's log K_{ow} value, the greater the root uptake. This can be demonstrated even on macerated root tissue. Root absorption of lipophilic organic compounds and subsequent root harvesting for processing and destruction may provide a useful remediation technique in specific instances (McMullin 1993); however, lipophilicity implies an equal propensity to partition into soil humic matter and onto soil surfaces. Root absorption and harvesting become increasingly difficult with heavily textured soils and soils higher in native organic matter. With compounds having lower log K_{ow} values, plant uptake is dependent to a larger degree on active plant processes, especially water transpiration rates. Plants that have high water processing capacities take up more of these types of compounds than do plants with reduced water requirements. Water transpiration fluxes are measured or can be calculated from water use efficiencies (WUE) multiplied by biomass produced. WUE values vary considerably, with alfalfa requiring 858 lb (389 kg) water transpired per pound (0.45 kg) of biomass produced, whereas sorghum requires only 271 lb (123 kg) (Aldrich et al. 1975).

In targeting phytoremediation of specific compounds, perhaps the best chemical model to use is that of a "preplant-incorporated herbicide." These are compounds that are mixed into soils, move through the soil solution or gas phase, get into plant roots, and interact with plant metabolism. If one views all of surficial soil pollutants as preplant-incorporated xenobiotics (because, in most cases priority pollutants are not toxic to plants at rates 10- to 100-fold higher than regulatory limits), these design criteria used in the development of new pesticides provide an excellent perspective of what compounds are likely to be remediated by plant-based systems. In general, for a compound in soil to be a good candidate for plant remediation it will have a log K_{ow} from 1.5 to 4. Compounds more lipophilic than this are too tightly bound in soil and are not available for movement in the soil solution phase (Bell 1992). Some lipophilic compounds with higher K_{ow} values can still be effective soil herbicides and still be targeted for plant uptake, but these compounds may need to have significant volatility, thereby traveling in the soil gas phase instead of the soil solution to the root surface. All preplant-incorporated pesticides (herbicides as well as fungicides) on the market today follow these physical/chemical principles.

Once the pollutant has been absorbed by the plant roots, it can have three fates. It may be (1) sequestered in the root tissue often in forms that are not accessible to solvent extraction, (2) metabolized in the root, and (3) transported out of the root and into the shoot either as parent material or in a slightly modified form. Sequestration of many xenobiotics is a common fate, often associated with cell wall materials such as lignin. Plants have significant metabolic activities both in the root and in the shoot. Many of these metabolic capacities tend to be enzymatically and chemically similar to those processes that occur in mammalian livers. Many reviews of plant metabolic behaviors have been published (Hathaway 1989). One author has even equated plants to "green livers" due to the similarities in detoxification processes where hydroxylation followed by glycosylation activities are common ways to rid an organism of lipophilic toxins (Sandermann 1992).

Significant differences also exist between the metabolic capacities of plants. These differences allow the development of the multibillion dollar selective herbicide industry. Selective herbicides are applied to weeds and crops alike. The vast majority of herbicidal compounds has been selected so that the crop species are capable of metabolizing the pesticide to nontoxic compounds, whereas the weed species either lack this capacity or accomplish the metabolism at too slow a rate. The result is the death of the weed species without the metabolic mechanism to rid itself of the toxin (Hatzios & Hoagland 1989). In reality, it is difficult to find compounds that act in such a manner. Tens of thousands of compounds are screened to obtain one compound where the crop species is tolerant and multiple weed species are not. Results from the extensive herbicide discovery efforts show that many of the most noxious weed species are considerably more tolerant to a wide array of toxins than are domesticated agronomic plants. In our many surveys of plant life on hazardous waste sites, we often find the same weed species that are targeted in the development of new herbicides. It is unfortunate that the literature database is primarily derived from pesticide applications on crop plants, whereas the literature base on both priority pollutants and weed metabolic activities is relatively meager by comparison. Obviously more work is needed in this area to advance the science of phytoremediation.

Some researchers are attempting to augment the inherent metabolic capacity of plants by incorporating bacterial, fungal, insect, and even mammalian genes into the plant genome. Plants are autotrophs, with generally more limited destructive capacities than heterotrophic organisms. In certain instances, but by no means all, increasing the metabolic capacity of plants through this mechanism holds promise. In certain cases the research is being carried out as an extension of molecular tools developed for other purposes. This research often tends to be driven by goals chosen because they can be readily accomplished with little forethought given as to the location and concentration of xenobiotic in the plant-root-soil system relative to gene expression and enzyme function. Targeting compounds that can get into the plant and are translocated in the plant is more appropriate. Targeting gene expression to the appropriate tissue has also not always been well considered. Lipophilic compounds that remain in the soil are perhaps best targeted by activities that occur at the soil-root surface interface and not those that occur internal to the plant itself.

Another potential source of increased metabolic capacities in plant-based systems is microbial communities that live both internal and external to the plant. Plants maintain active microbial communities around their roots, in their root tissue, in the xylem stream, in shoot and leaf tissue, and on the surface of leaves. Perhaps the best studied, and most active, microbial communities exist in the root zone. These rhizosphere communities have been the subject of much recent focus on the phytoremediation of chlorinated solvents and petroleum hydrocarbons. To date, most research has been preliminary and suggestive rather than producing definitive results on the capacity of these communities to remediate these contaminants. Evidence of accelerated xenobiotic degradation has been obtained, for example, trichloroethylene (Anderson et al. 1993) and petroleum hydrocarbon degradation (Aprill & Sims 1990), but the rate has been slow and the final quantity degraded has been low by remediation engineering standards. The technology has been reviewed recently (Anderson & Coats 1994). This is a new and evolving area of phytoremediation, with few remediation successes reported to date. The concept is still primarily in the lab and greenhouse stages of development. "Rhizodegradation" builds on a solid microbial and plant science foundation, but real field validation is still a few years away. There are perhaps a dozen field tests scheduled for spring '95, targeting soils contaminated with petroleum hydrocarbons, pesticide spills, and chlorinated solvents. For the most part, these are early attempts at choosing plants, agronomic practices, and soil amendments for phytoremediation. More sophisticated approaches, crops, and cultural techniques will be forthcoming as research results from many labs begin to be reported.

FUTURE STATE

Modern molecular biology and plant biology have begun to address some of the inherent limitations of current plants for phytoremediation. The agricultural industry and molecular biologists have developed many tools over the last decade that, although initially discarded, may be suitable for the development of new plants for phytoremediation. Many of the older genetic clones, promoters, and genes had significant negative impacts on crop yield. This was deemed unacceptable and many such projects have since been redirected. In phytoremediation, however, agronomic yield is not a driving factor. For phytoremediation purposes, these genetic tools and plants may not only be acceptable, but, in some instances, preferable to high biomass phenotypes. Plants used in either phytostabilization of inorganics or phytoremediation of organics may not need high shoot biomass, and reduced harvesting schedules may be economically and logistically beneficial. The engineering community is often sensitive to fears over the use of recombinant plants. Traditional plant breeding may provide these tools as well, albeit probably at a slower rate. There is increasing evidence, however, to suggest that transgenic plants may be becoming broadly acceptable, especially as a tool to remediate environmentally damaged areas. Field tests of transgenic plants created for agronomic reasons are now common with more than 1,000 field trials conducted during 1994. Plants designed for remediation purposes are emerging in many labs. Plants with altered rooting structures (Gordon et al. 1989), differential metabolism (commercially available Basta® resistance), and increased metal binding (Misra & Gedamu 1990) are emerging from laboratories. Field tests with recombinant plants at contaminated sites are still in the planning stages.

One of the most exciting prospects is to combine phytoremediation with traditional engineering remediation techniques. Many such combinations are being tried in laboratories throughout the world. One such example from our lab is to combine the process of electroosmosis with phytoextraction. Relatively immobile cations can be coaxed to migrate by the application of an electrical direct current across the soil. Generally, voltages are applied over long periods of time and require solution-handling techniques at one or both electrodes to maintain pH balances and remove contaminants. Plant roots have special distributions in the soil profile. There is a significant kinetic limitation to metal uptake for relatively immobile ions located at some distance from the nearest root. By applying a direct current, and alternating its direction at regular intervals, we can increase the rate of contaminant migration to the root and significantly increase plant loading rates (manuscript in preparation). Using the plant root as a sink, rather than electrodes spaced at relatively large intervals, may further eliminate the need for solution-handling systems at the electrode(s), allowing for increased electrode spacing and decreased time and power requirements of traditional electromigration methods. Other such hybrid technologies look as technically, economically, and scientifically promising.

CONCLUSIONS

Phytoremediation is an emerging technology that holds a promising future. Plant-based systems have physical, chemical, and biological processes that tend to ameliorate a wide variety of environmental contaminants and condition the soil for continued plant growth. Some inorganic contaminants can be stabilized in place, with documented reduction in environmental hazard. Others can be drawn into plant tissue for harvesting and reclamation, although neither the rate nor the quantity of the latter process has been optimized. Organic molecules can be similarly sequestered, absorbed, and translocated, but in addition, can be degraded either by the plant or plant-associated microflora. Analytical techniques are a critical factor in both the development and acceptability of phytoremediation. It is doubtful that a plant-based system can ever remediate a soil to treatment endpoints that are derived from incineration techniques or based on total acid digestion. As in microbial bioremediation schemes, contaminants can be remediated only if they are bioavailable. Organic molecules sequestered into nano-sized pores in the soil and metals bound in the interstices of clay lattices are not mobile, nor are they available to plant roots and microbes. To the extent that regulatory processes acknowledge that bioavailability of contaminants is correlated to biological hazard, phytoremediation has a promising future. The technical, economic, and regulatory climate for continued development of phytoremediation is excellent.

REFERENCES

Aldrich, S. R., W. O. Scott, and E. R. Leng. 1975. *Modern Corn Production*, 2nd ed. A & L Publishers, Champaign, IL.

Alexander, M. 1994. *Biodegradation and Bioremediation*. Academic Press, San Diego, CA.

Anderson, T. A., and J. E. Coats (Eds.). 1994. *Bioremediation through Rhizosphere Technology*. ACS Symposium Series 563. American Chemical Society, Washington, DC.

Anderson, T. A., E. A. Guthrie, and B. T. Walton. 1993. "Bioremediation in the Rhizosphere." *Environmental Science and Technology, 27*: 2630-2635.

Aprill, W., and R. C. Sims. 1990. "Evaluation of the Use of Prairie Grasses for Stimulating Polycyclic Aromatic Hydrocarbon Treatment in Soil." *Chemosphere 20*: 253-265.

Baker, A.J.M., and R. R. Brooks. 1989. "Terrestrial Higher Plants which Hyperaccumulate Metallic Elements — A Review of their Distribution, Ecology, and Phytochemistry." *Biorecovery 1*: 81-126.

Baker, A.J.M., R. D. Reeves, and S. P. McGrath. 1991. "In Situ Decontamination of Heavy Metal Polluted Soils Using Crops of Metal-Accumulating Plants — A Feasibility Study." In R. E. Hinchee and R. F. Olfenbuttel (Eds.), *In Situ Bioreclamation, Applications and Investigations for Hydrocarbon and Contaminated Site Remediation*, pp. 600-605. Butterworth-Heinemann. Stoneham, MA.

Bell, R. M. 1992. *Higher Plant Accumulation of Organic Pollutants from Soils*. U.S. Environmental Protection Agency Technical Report. EPA 600/R-92/138, Risk Reduction Engineering Laboratory, ORD, Washington, DC.

Berti, W. R., and S. D. Cunningham. 1995. "Remediating Soil Pb with Green Plants." In *Trace Substances, Environment and Health*. Science Reviews, Northwood, UK (in press).

Bradshaw, A. D., and M. D. Chadwick. 1980. *The Restoration of Land: The Geology and Reclamation of Derelict and Degraded Land*. University of California Press, Berkeley/Los Angeles, CA.

Brandon, D. L., C. R. Lee, J. W. Simmers, and J. G. Skogerboe. 1991. *Long Term Evaluation of Plants and Animal Colonizing Contaminated Estuarine Dredged Material Placed in Upland and Wetland Environments*. Misc. Paper D-91-5-US, Army Engineer Waterways Experimental Station, Vicksburg, MS.

Brooks, R. R., J. Lee, R. D. Reeves, and T. Jaffre. 1977. "Detection of Nickeliferous Rocks by Analysis of Herbarium Specimens of Indicator Plants." *Journal of Geochemical Exploration, 7*: 49-77.

Brown, S. L., R. L. Chaney, J. S. Angle, and J. M. Baker. 1994. "Phytoremediation Potential of *Thlaspi caeurlescens* and Bladder Campion for Zinc- and Cadmium-Contaminated Soil." *Journal of Environmental Quality, 23*: 1151-1157.

Cunningham, S. D., and W. R. Berti. 1993. "Remediation of Contaminated Soils with Green Plants: An Overview." *In Vitro Cellular and Developmental Biology, 20P*: 207-212.

Delhaize, E., P. J. Randall, P. A. Wallace, and A. Pinkerton. 1994. "Screening *Arabidopsis* for Mutants in Mineral Nutrition." *Plant Soil, 156*: 131-134.

Gordon I., S. A. Sojka, and M. P. Gordon. 1989. U.S. Patent No. 369886.

Grusak, M. A. 1994. "Iron Transport to Developing Ovules of *Pisum sativum* I: Seed Import Characteristics and Phloem Iron-Loading Capacity of Source Regions." *Plant Physiology 104*: 649-655.

Hathaway, D. E. 1989. *Molecular Mechanism of Herbicide Selectivity.* Oxford University Press, New York, NY.

Hatzios K. K., and R. E. Hoagland. 1989. *Crop Safeners for Herbicides: Development, Uses and Mechanism of Action.* Academic Press, New York, NY.

Huang, J. W., D. L. Grunes, and L. V. Kochian. 1993. "Aluminum Effects on Calcium Uptake and Translocation in Wheat Forages." *Agron. J., 85*:867-873.

Jaffre, T. 1979. "Accumulation du Manganese par les Protéacées de Nouvelle-Caledonie," *C.R. Acad. Sci. Ser. D* (Paris) *289*: 425.

Jaffre, T. 1980. "Etude Ecologique du Peuplement Vegetal des Sols Derivés des Roches Ultrabasiques en Nouvelle-Caledonie." *Travaux et Documents de 1' O.R.S.T.O.M No. 124*, Paris.

Jaffre, T., R. R. Brooks, J. Lee, and R. D. Reeves. 1976. "*Sebertia acuminata*: A Hyperaccumulator of Nickel from New Caledonia." *Science, 193*: 579-580.

Jeffery, D. W. 1987. *Soil-Plant Relationships: An Ecological Approach.* Timber Press, Portland, OR.

Malaisse, F. F., F. Gregoire, R. R. Brooks, and R. D. Reeves. 1979. "Copper and Cobalt Vegetation in Fungurume, Shaba Province, Zaire." *Oikos 33*: 472-478.

McMullin, E. 1993. "Absorbing Idea." *California Farmer Feb 1993*: 20-24.

Misra, S., and L. Gedamu. 1990. "Heavy Metal Resistance in Transgenic Plants Expressing a Human Metallothionein Gene." *Plant Gene Transfer*. Alan R. Liss Inc., New York, NY.

Pignatello, J. J. 1989. "Sorption Dynamics of Organic Compounds in Soils and Sediments." In B. L. Sawhney, and K. Brown (Eds.), *Reactions and Movements of Organic Chemical in Soil*, pp 45-80. Soil Science Society of America Special Publication #22, Madison, WI.

Reeves, R. D., and R. R. Brooks. 1983a. "European Species of *Thlaspi* L. (Cruciferae) as Indicators of Nickel and Zinc." *Journal of Geochemical Exploration 18*: 275-283.

Reeves, R. D., and R. R. Brooks. 1983b. "Hyperaccumulation of Lead and Zinc by Two Metallophytes from Mining Area of Central Europe." *Environ Pollution Ser. A., 31*: 277-285.

Ruby, M. V., A. Davis, T. W. Link, R. Schoof, R. L. Chaney, G. B. Freeman, and P. Bergstrom. 1993. "Development of an In Vitro Screening Test to Evaluate the In Vivo Bioaccessibility of Ingested Mine-Waste Lead." *Environmental Science and Technology 27*(13): 2870-2877.

Sandermann, H., Jr. 1992. "Plant Metabolism of Xenobiotics." *Trends in Biotechnology 17*: 82-84.

Welch, R. M., and T. A. LaRue. 1990. Physiological Characteristics of Fe Accumulation in the 'Bronze' Mutant of *Pisum sativum* L., cv 'Sparkle' E107 (brz brz)." *Plant Physiology 93*: 723-729.

Phytoremediation of Soils Contaminated with Toxic Elements and Radionuclides

James E. Cornish, William C. Goldberg,
Rashalee S. Levine, and John R. Benemann

ABSTRACT

At many U.S. Department of Energy (U.S. DOE) facilities and other sites, surface soils over relatively large areas are contaminated with heavy metals, radionuclides, and other toxic elements, often at only a relatively small factor above regulatory action levels. Cleanup of such sites presents major challenges, because currently available soil remediation technologies can be very expensive. In response, the U.S. DOE's Office of Technology Development, through the Western Environmental Technology Office, is sponsoring research in the area of phytoremediation. Phytoremediation is an emerging technology that uses higher plants to transfer toxic elements and radionuclides from surface soils into aboveground biomass. Some plants, termed "hyperaccumulators," take up toxic elements in substantial amounts, resulting in concentrations in aboveground biomass over 100 times those observed with conventional plants. After growth, the plant biomass is harvested, and the toxic elements are concentrated and reclaimed or disposed of. As growing, harvesting, and processing plant biomass is relatively inexpensive, phytoremediation can be a low-cost technology for remediation of extensive areas having lightly to moderately contaminated soils. This paper reviews the potential of hyper- and moderate accumulator plants in soil remediation, provides some comparative cost estimates, and outlines ongoing work initiated by the U.S. DOE.

INTRODUCTION

The U.S. Department of Energy (DOE) is faced with cleanup of heavy metal and radionuclide-contaminated soils and sediments at a number of facilities throughout the United States (U.S. DOE 1995). About half of the DOE facilities report near-surface soils contaminated with toxic and radioactive elements (e.g., As, Cd, Cu, Cr, Hg, and Zn, and isotopes of Cs, Pu, Sr, and U) (Riley et al. 1992).

It has been suggested (e.g., Chaney 1983; Baker et al. 1991; Cunningham and Berti 1994; Brown et al. 1994; Raskin et al. 1994) that higher plants, in particular the hyperaccumulator plants, could remove toxic elements, such as heavy metals, from soils. The general approach is to cultivate plants on the contaminated soils over a number of growth/harvest cycles to transfer the toxic elements into aboveground leaf and stalk biomass (LSB). The LSB is then removed and processed (e.g., ashed) to further reduce the final volume that must be disposed to as little as 1 one-thousandth of the mass of contaminated soil. Applications of this process to the U.S. DOE complex are reviewed herein.

HYPERACCUMULATOR PLANT SPECIES

Resistance to metals phytotoxicity can be facilitated either by exclusion of the metal(s) from the plant or via internal tolerance mechanism(s) (Shaw 1990). Hyperaccumulator plants are those that exhibit the ability to tolerate high concentrations of toxic metals in aboveground plant tissues; these species contain toxic element levels in the LSB of about 100 times those of nonaccumulator plants growing in the same soil, with some species and metal combinations exceeding conventional plant levels by a factor of more than a thousand (Baker and Brooks 1989). For soils contaminated with radionuclides, or low levels of toxic elements, even conventional plants, or "accumulator" species (e.g., those with only 10 times the normal metal levels in their LSB) could be useful in phytoremediation.

Table 1 lists the reported concentrations or radioactivity ranges for the most frequently reported inorganic CoCs (contaminants of concern) in soils and sediments at U.S. DOE facilities. Table 1 also estimates expected phytotoxicity thresholds for conventional plants in both soils and plant biomass, and the levels of such toxic elements in the LSB that should be exhibited by plants to qualify as hyperaccumulator species. Finally, Table 1 presents preliminary soil remediation goals, estimated to be both reasonably achievable and protective of human health and the environment. With the possible exception of Cu, all of the elements in Table 1 are present, at one or more DOE facilities, at levels exceeding their respective preliminary remediation goals.

There is a relatively large number of publications on the toxic metal and radionuclide content of plants, including hyperaccumulator species. However, relatively few of these studies are directly useful because in many, if not most, cases adequate soil metal content data are lacking. Also, many of the studies in which both soil and plant analyses are reported are for hydroponic or pot tests, which are often not representative of "real world" situations (deVries and Tiller 1978). With these provisos, some selected, and generally representative, literature on hyperaccumulator species is summarized in Table 2. Accumulation coefficients (AC, ratio of toxic element concentration in the biomass to that in soil) range from <1 to >10 in these studies. For phytoremediation to be successful, an initial AC of above 10, and preferably 40, is required to allow treatment of large amounts of soil by plant biomass in a reasonable period of time (<20 years).

TABLE 1. Contaminant-of-concern data utilized in evaluating the feasibility of phytoremediation of soils within the U.S. DOE Complex.[a][b]

	Soils			Plants	
Element	Concentration Range	Phytotox. Threshold	Preliminary Remedial Goals	Phytotox. Threshold	Hyperaccum. Species[c]
Part A. Heavy Metals (μg/g DW)					
As	0.1 to 102	50	25 to 50	20	100
Cd	0.1 to 345	50	2 to 5	25	100
Cr (total)	<0.1 to 3,950	500	500 to 1,000 ($Cr^{+6} \leq 50$)	15	1,000
Cu	<0.1 to 550	500	250 to 500	25	1,000
Hg (inorg.)	<0.001 to 1,800	5	1 to 3	3	10
Pb	1 to 6,900	1,000	500 to 1,000	35	1,000
Zn	<0.2 to 5,000	1,000	1,000 to 1,500	150	10,000
Part B. Radionuclides (pCi/g DW)					
Radio-nuclide(s)	**Activity Range**	**Phytotox. Threshold**	**Preliminary Remedial Goals**	**Phytotox. Threshold**	**Hyperaccum. Species**[c]
^{137}Cs	2×10^{-5} to 46.9	ND[d]	1.0	1×10^{7}	20
^{239}Pu	1×10^{-7} to 3,500	ND[d]	0.04	ND[d]	1.0
^{90}Sr	$<1\times10^{-4}$ to 540	ND[d]	0.5	5×10^{6}	2.5
$^{238,234}U$ (natural)	<0.1 to 18,700	2,000	35	1×10^{4}	68 (~100 ppm)

(a) Concentration and radioactivity data are from Table 7 in Riley et al. (1992).
(b) All other data are estimates based on literature reviws and best judgments by the authors.
(c) Defined as those vascular plants that contain 100-fold greater amount of a particular element or radionuclide than is typically observed in air- or oven-dried (DW) biomass in the given species.
(d) ND means not determined.

Table 2 indicates the existence of herbaceous and shrubby species that can hyperaccumulate one or more of the listed heavy metals. Table 2 also suggests that the phytoremediation of radionuclides such as Cs and Sr appears possible, without the need for hyperaccumulating plants. However, with the possible (though at present uncertain) exception of U, the actinides do not appear to be capable of accumulation by plants (Trabalka and Garten 1983).

RELATIVE ECONOMICS OF PHYTOREMEDIATION

The potential for cost-effective phytoremediation of sites within the DOE complex, compared to conventional technologies, is illustrated by the following

TABLE 2. Contaminant-specific herbaceous, hyperaccumulator plant species — heavy metals and radionuclides.[a]

Element	Plant Species (family)	Accum. Coeff. (or conc. ratio)[b]	Reference	Maximum Observed	
				Plants[c]	Soils[c]
As	*Agrostis tenuis* Sibth. (Poaceae)	0.2 (ND)	Porter and Peterson (1977)	3,460	17,220[d]
Cd	*Thlaspi caerulescens* J. and C. Presl (Brassicaceae)	1	Brown et al. (1994)	1,020	1,020
Cu	*Aeollanthus subacaulis* var. *linearis* (Burk.) Ryding (Lamiaceae)	(ND)	Brooks et al. (1992)	13,700[e]	
Pb	*Thlaspi rotundifolium* (L) Gaudin *cepaeifolium* (Wulf.) Rouy et Fouc. (Brassicaceae)	2.7 (ND)	Reeves and Brooks (1983)	8,200	3,000[d]
Zn	*Thlaspi caerulescens* J. and C. Presl (Brassicaceae)	10	Brown (1993)	30,000	3,000
^{137}Cs	*Paspalum notatum* Flugge cv. saurae Parodi (Poaceae)	1.56 to 9.8	Adriano et al. (1984)	2,810	285
^{90}Sr	*Arabis stricta*, no attribution cited by authors (Brassicaceae)	21 (ND)	Bowen and Dymond (1956)	10,600	500[d]
U	*Uncinia leptostachya*, no attribution cited by authors (Cyperaceae)	0.7	Whitehead and Brooks (1969)	25,100	36,500

(a) Selected from a database developed by J. Cornish (1994, unpublished).
(b) Calculated from reported data or by the authors of the original reports. (ND) means not disclosed in the references cited.
(c) Values for nonradioactive elements are given in µg/g, while those for radioactive isotopes are given in pCi/g.
(d) Denotes mean value.
(e) Denotes dried plant material.

hypothetical example. The site (assumed to be located in the western United States) is a 0.5-ha chemical waste disposal pond that received Cd, Zn, and ^{137}Cs process effluents over several decades. Because phytoremediation is generally restricted to the plant rooting zone, this example assumes the contaminants are concentrated in the top 50 cm of soil. The maximum contaminant levels and respective remediation goals (in parentheses) for "clean closure" for this site are 75 (5) mg/kg of Cd, 2,580 (1,500) mg/kg of Zn, and 22 (1) pCi/g of ^{137}Cs. For phytoremediation of this site, a mixture of *Thalspi caerulescens* and *Paspalum notatum* was selected, based on the hyperaccumulation properties shown in Table 2.

The site is tilled with conventional agricultural equipment in late spring, and then seeded with an equal mix of these plants. A total dry biomass productivity of 10 tons hectare^{-1} year^{-1} (t/ha/y) is assumed for each of the plants, achieved by adding N-P-K fertilizer and using a drip or similar irrigation system, as needed. At the end of the growing season, the "meadow" is cut and baled, again with conventional agricultural equipment. The plant biomass is field dried, shredded, and ashed using a low-temperature ashing process (to be selected from among a relatively large number of different available technologies, Rosholt et al. 1994). The process ash and air pollution control system residues are collected for disposal in a hazardous waste landfill.

The next major issue is the time required for cleaning up the contaminated soil. For the present example, it is assumed, for lack of a better model at present, that the AC is a constant (regardless of soil contamination levels) as phytoremediation proceeds, a probably conservative assumption. With this model, the equation for estimating the final concentration after a certain number of years of phytoremediation is $C_t = C_0 e^{-kt}$, with C_t being the soil concentration after a given time period, t (years), of phytoremediation; C_0 being the initial soil concentration; and k being the product of the plant mass/soil ratio (20 t/ha/y/7,100 t/ha for the top 50 cm, assuming a bulk density of 1,420 kg/m^3 for the soil) times the AC, assumed to be 10 in this example. With these assumptions, after 20 years of phytoremediation, the final soil concentrations of these contaminants would be 4.5 mg/kg of Cd, 214 mg/kg of Zn, and 1.3 pCi/g of ^{137}Cs. The last is slightly above cleanup goals, but should be acceptable for such a site.

Based on the above assumptions, a preliminary cost estimate was calculated and compared to the cost for a conventional soil washing process (Table 3). Although both cost estimates are somewhat speculative, due to present uncertainties in both cleanup technologies, the three-fold lower projected cost for phytoremediation is considered reasonable, because plant cultivation requires far less materials handling than soil excavation and processing. Furthermore, the overall costs for phytoremediation can be spread over a relatively long period of time.

The above example is probably conservative: the assumed accumulation factors are relatively low, and, most important, the site to be treated quite small. Cost comparisons would be even more favorable for larger sites, where phytoremediation would exhibit significant economies of scale, compared to soil washing technologies. Indeed, it is for relatively large (>1 ha) sites that phytoremediation

TABLE 3. Screening level cost estimates for remeditaion of the (hypothetical) chemical waste disposal pond sediments by phytoremediation versus soil washing.[a]

Activity	Cost Estimate ($)		Summary of Activity Components
	Phytoremediation[b]	Soil Washing	
Site preparations	63,000	61,000	Surveying, staking, and light clearing of research area, access construction, excavations, installation of water supplies, and construction of fences and other structures.
Production/processing	395,000	1,336,000	Operation and maintenance of treatment system for the duration of the project, including heavy equipment rentals and services.
Waste management/disposal	35,000	175,000	Consolidation of waste materials and haulage of same to eventual storage facility, as well as necessary testing prior to disposal. Costs for long-term storage have not been calculated due to uncertainties regarding the ultimate deposition of some wastes.
Project closeout	18,000	62,000	Demobilization of heavy equipment, removal of constructed facilities, and costs for site restoration.
Environmental, safety & health (ES&H), engineering design (ED), and project management overhead (PMO)	153,000	490,000	Calculated as 30% of above costs.
Overall estimated cost	664,000	2,124,000	Subtotal plus ES&H, ED, and PMO costs.
Cost per tonne[c] of sediment	200	600	Based on treating 2,500 m^3 of sediment @ 1,420 kg/m^3.

(a) Details regarding these estimates are contained in MSE project files; key references include Smit et al. (1994); Rosholt et al. (1994); Anderson (1993); and Gerber et al. (1991).
(b) Costs shown for phytoremediation assume a production period of 20 years.
(c) Denotes units of metric tons (1,000 kg per tonne).

would be most applicable. However, it must also be recognized that phytoremediation must still be demonstrated in the field. That is the objective of ongoing research supported by the U.S. DOE, briefly described below.

ONGOING U.S. DOE-SPONSORED R&D IN PHYTOREMEDIATION

Phytoremediation of contaminated soils is an active area of research at a number of laboratories in the United States and abroad (Benemann et al. 1994; Parry 1995) with several private companies entering this field (e.g., Phytotech, see Atlas 1995; Dupont, see Cunningham and Berti 1994). The U.S. DOE's Office of Technology Development, through the Western Environmental Technology Office (WETO) in Butte, Montana, has initiated a broad-based program to evaluate and develop phytoremediation technologies. An extensive database on hyperaccumulators has been developed (a small fraction of which was presented in Table 2); alternative biomass processing (volume reduction) technologies have been evaluated on a preliminary basis (Rosholt et al. 1994), and the relative economics of phytoremediation assessed, as reviewed above. Bench-scale research at WETO evaluated a "Fractionation Separation Technology" for radionuclide-contaminated biomass, and field studies screened several dozen native plant species for phytoremediation potential at Zn- and Cd-contaminated sites in Montana (Cornish et al. 1994).

Screening for U-accumulating plants was initiated in the spring of 1994 at the U.S. DOE's Fernald, Ohio site and at an abandoned uranium mine site near Clancy, Montana. Heavy metal accumulator plant screening was initiated also that spring at a test plot located near Anaconda, Montana. This search was expanded during late summer 1994 to include the sampling of LSB from indigenous vegetation and associated soils at abandoned hardrock mine and mill sites located in southwestern Montana and northern Idaho, or near Leadville, Colorado.

Modest uptake of U (i.e., up to 10 mg U/kg LSB) was observed in commercially available red clover (*Trifolium repens* cv. "medium red") at Fernald and in tansy mustard (*Descurainia pinnata*) at Clancy (Cornish et al. 1995). A number of herbaceous species sampled at the mine and mill sites appear to be metals accumulators and/or metals tolerant. The concentrations of metals in LSB were often at "background" levels (e.g., ≤ 1 mg Cd/kg), although concentrations ≥ 10 times background were occasionally observed; in all of these instances, acid-extractable and plant-available metals levels in the "soils" were judged to be phytotoxic (based on MSE's literature review). An interesting case of metal tolerance and accumulation was noted at the Independence mill tailings site in northern Idaho. Acid extractable and plant available Zn levels in the "soils" were 11,900 mg/kg and 2,630 mg/kg, respectively. Colonial bentgrass (*Agrostis capillaris* cv. Parys) and red fescue (*Festuca rubra* cv. Merlin) used to reclaim this site contained 3,770 mg Zn and 2,480 mg Zn per kg LSB, respectively (report in preparation).

The U.S. Department of Agriculture's Plant Soil and Nutrition Laboratory at Ithaca, New York is currently screening hundreds of plant accessions for capability of hyperaccumulating certain radionuclides (i.e., ^{137}Cs, ^{90}Sr, and U) and/or nonradioactive elements (e.g., Cd, Zn). The hydroponic solutions being used were developed to simulate the soil solution chemistry present at the test sites to be evaluated (for phytoremediation feasibility) this coming summer.

Preliminary cost/benefit analyses and risk analysis/management assessments will be carried out to support this research and evaluate the site-specific applications of phytoremediation. Although phytoremediation is still at an early stage in technological development, both fundamental and applied knowledge in this field suggest that this approach could solve many of the most difficult soil cleanup problems at U.S. DOE sites, and around the world.

ACKNOWLEDGMENTS

This work was funded by the Office of Technology Development (OTD), within the U.S. DOE Office of Environmental Management; it was performed under prime contract DE-AC22-88ID-12735 between the U.S. DOE's OTD and MSE, Inc., Butte, Montana.

REFERENCES

Adriano, D. C., K. W. McLeod, and T. G. Ciravolo. 1984. "Long-term Root Uptake of Radiocesium by Several Crops." *Journal of Plant Nutrition* 7(10): 1415-1432.

Anderson, W. C. (Ed.). 1993. *Soil Washing/Soil Flushing*. Vol. 1 of the Innovative Site Remediation Technology Series, American Academy of Environmental Engineers, Annapolis, MD.

Atlas, R. M. 1995. "Special Report: Bioremediation." *Chemical and Engineering News* 73(14): 32-42.

Baker, A.J.M., and R. R. Brooks. 1989. "Terrestrial Higher Plants which Hyperaccumulate Metallic Elements — A Review of their Distribution, Ecology and Phytochemistry." *Biorecovery* 1: 81-126.

Baker, A.J.M., R. D. Reeves, and S. P. McGrath. 1991. "In Situ Decontamination of Heavy Metal Polluted Soils using Crops of Metal-Accumulating Plants – A Feasibility Study." In R. E. Hinchee and R. F. Olfenbuttel (Eds.), *In Situ Bioreclamation: Applications and Investigations for Hydrocarbon and Contaminated Site Remediation*, pp. 600-605. Butterworth-Heinemann, Stoneham, MA.

Benemann, J. R., R. Rabson, and R. S. Levine. 1994. "Summary Report of a Workshop on Phytoremediation Research Needs, July 24-26, 1994, Santa Rosa, California" (Draft). U.S. Department of Energy, Offices of Technology Development and Energy Biosciences, Germantown, MD.

Bowen, H.J.M., and J. A. Dymond. 1956. "Strontium and Barium in Plants and Soils." *Proceedings of the Royal Society of London, B 144*: 355-368.

Brooks, R. R., A.J.M. Baker, and F. Malaisse. 1992. "Copper Flowers." *National Geographic Research and Exploration* 8(3): 338-351.

Brown, S. L. 1993. "Zinc and Cadmium Uptake by the Metal Tolerant Plants, *Thlaspi alpestre* L. and *Silene cucubalis* L." M.S. Thesis, University of Maryland, College Park, MD.

Brown, S. L., R. L. Chaney, J. S. Angle, and A.J.M. Baker. 1994. "Phytoremediation Potential of Thlaspi caerulescens and Bladder Campion for Zinc- and Cadmium-Contaminated Soil." *Journal of Environmental Quality* 23(6): 1151-1157.

Chaney, R. L. 1983. *Land Treatment of Hazardous Wastes*. Part 1, Chapter 3. Noyes Data Corporation, Park Ridge, NJ.

Cornish, J., W. Goldberg, and R. Levine. 1995. "Evaluation of In Situ Phytoremediation of Uranium-Contaminated Soils in Ohio and Montana." Presented at the Waste Management '95 Conference, March 1995, Tucson, AZ.

Cornish, J. E., D. Alexander, T. Erickson, W. Goldberg, S. Kujawa, and S. Larson. 1994. "Proof of Concept Testing of MTI-Biotech's Biomass Remediation System (BRS): Phase 1 (August-September 1993) Test Results" (Internal Review Draft). Prepared for the U.S. DOE, Office of Technology Development by MSE, Inc.

Cunningham, S. D., and W. R. Berti. 1994. "Remediation of Contaminated Soils with Green Plants: An Overview." *In Vitro Cell. Dev. Biol. 29P*: 207-212.

deVries, M.P.C., and K. G. Tiller. 1978. "Sewage Sludge as a Soil Amendment, with Special Reference to Cd,Cu, Mn, Ni, Pb, and Zn — Comparison of Results from Experiments Conducted Inside and Outside a Glasshouse." *Environmental Pollution 16*: 231-240.

Gerber, M. A., H. D. Freeman, E. G. Baker, and W. F. Riemath. 1991. *Soil Washing: A Preliminary Assessment of its Applicability to Hanford*. PNL-7787/UC-902. Prepared for the U.S. DOE by Pacific Northwest Laboratory, Richland, WA.

Parry, J. 1995. "Plants Absorb Heavy Metals." *Pollution Engineering* 27(2): 40-41.

Porter, E. K., and P. J. Peterson. 1977. "Arsenic Tolerance in Grasses Growing on Mine Waste." *Environmental Pollution 14*: 255-265.

Raskin, I., PBA N. Kumar, S. Dushenkov, and D. E. Salt. 1994. "Bioconcentration of Heavy Metals by Plants." *Current Opinion in Biotechnology 5*: 285-290.

Reeves, R. D., and R. R. Brooks. 1983. "Hyperaccumulation of Lead and Zinc by Two Metallophytes from Mining Areas of Central Europe." *Environmental Pollution (Series A) 31*: 277-285.

Riley, R. G., J. M. Zachara, and F. J. Wobber. 1992. *Chemical Contaminants on DOE Lands and Selection of Contaminant Mixtures for Subsurface Science Research*. DOE/ER-0547T. Prepared for U.S. DOE by Pacific Northwest Laboratory, Richland, WA.

Rosholt, D. L., J. Modrell, D. Lodman, B. Archibald, J. Baughman, J. Cornish, and W. Goldberg. 1994. "Biomass Remediation System (BRS): Volume Reduction of Heavy Metal and Radionuclide Hyper-Accumulating Biomass Technology Ranking Report" (Internal Review Draft). Prepared for U.S. DOE's Office of Technology Development by MSE, Inc.

Shaw, A. J. (Ed.). 1990. *Heavy Metal Tolerance in Plants: Evolutionary Aspects*. CRC Press, Inc., Boca Raton, FL.

Smit, K., H. M. Chandler, J. H. Chiang, K. Foley, P. L. Jackson, A. E. Lew, R. W. Mewis, M. J. Mossman, and others (Eds.). 1994. *Means Heavy Construction Cost Data*, 8th annual ed. R.S. Means Co., Inc., Kingston, MA.

Trabalka, J. R., and C. T. Garten, Jr. 1983. "Behavior of the Long-Lived Synthetic Elements and Their Natural Analogs in Food Chains." *Advances in Radiation Biology 10*: 39-104.

U.S. Department of Energy, Office of Environmental Management. 1995. *Closing the Circle on the Splitting of the Atom: The Environmental Legacy of Nuclear Weapons Production in the United States and What the Department of Energy is Doing About It*. Office of Strategic Planning and Analysis (EM-4), Washington, DC.

Whitehead, N. E., and R. R. Brooks. 1969. "Radioecological Observations on Plants of the Lower Buller Gorge Region of New Zealand and their Significance for Biogeochemical Prospecting." *Journal of Applied Ecology 6*: 301-310.

Bioremoval of Toxic Elements With Aquatic Plants and Algae

Tsen C. Wang, Joseph C. Weissman, Geetha Ramesh, Ramesh Varadarajan, and John R. Benemann

ABSTRACT

Aquatic plants were screened to evaluate their ability to adsorb dissolved metals. The plants screened included those that are naturally immobilized (attached algae and rooted plants) and those that could be easily separated from suspension (filamentous microalgae, macroalgae, and floating plants). Two plants were observed to have high adsorption capabilities (mg metal/kg of biomass) for cadmium (Cd) and zinc (Zn) removal: one blue green filamentous alga of the genus *Phormidium* and one aquatic rooted plant, water milfoil (*Myriophyllum spicatum*). These plants could also reduce the residual metal concentration to 0.1 mg/L or less. Both plants also exhibited high specific adsorption for other metals (Pb, Ni, and Cu) both individually and in combination. Metal concentrations were analyzed with an atomic absorption spectrophotometer (AAS).

INTRODUCTION

Aquatic plants and algae are known to accumulate metals and other toxic elements from solution (Wolverton and McDonald 1975, Muramoto and Ohi 1983, Sela et al. 1990, Chawla et al. 1991, Guven et al. 1992, Green and Bedell 1990, Wilde and Benemann 1993). There are large differences in bioremoval due to species and strain differences, cultivation methodology, and experimental protocols. Some systems have been shown to operate at low pH and may be applicable to the treatment or polishing of acid mine drainage waters (Stevens et al. 1989, Nakatsu and Hutchinson 1988). Commercial systems used immobilized algal biomass for removing U and Hg and other metal ions from wastewaters (Darnall et al. 1986, Green and Bedell 1990, Feiler and Darnall 1991). Macroalgae also appear to have potential in bioremoval. However, considerable limitations to commercial application still exist. The commercially available biomass sources used in most studies are generally not satisfactory for specific bioremoval applications. Therefore, a screening program, using simulated and actual contaminated waters, is a required first step in development of bioremoval processes. The goal of this study was to screen for promising plant and algae species for such a process. A successful candidate for bioremoval must bind large quantities of metal while reducing residual metal concentrations to a low level and be suitable for mass cultivation.

METHODOLOGY

Plants Used

The plants used in the screening experiments included six different vascular plants, two macroalgae (including four different strains of one seaweed genus), and five different microalgae. *Gracillaria Strain G16* (red, brown, green), *Ulva lactuca*, and *Lemna minor* (duckweed) were collected from the Aquaculture Division of Harbor Branch Oceanographic Institution (HBOI). *Hydrilla verticillata* (hydrilla), *Pistia stratiotes* (water lettuce), *Myriophyllum spicatum* (water milfoil), *Hygrophila polysperma* (hygrophyllum), and *Alternanthera philoxeroides* (alligator weed) were obtained from the University of Florida, Institute of Food and Agricultural Sciences, Agricultural Research Center, Ft. Lauderdale, Florida. One wild strain of duckweed and *Gracillaria* were collected from local growing areas. *Spirulina* sp. was provided by Dr. J. Cysewski, Cyanotech Corp. in Hawaii and algal turf sample was from Prof. W. Oswald, University of California, Berkeley. *Phormidium*, *Synechococcus*, and *Nitzchia* were collected from the Microbial Products, Inc., culture collection. The plants collected were brought to the laboratory and washed with deionized water to remove surface contamination.

In the laboratory, 50 mL of known metal solution was added to a 125-mL conical flask equipped with a screw cap. The required amount of biomass was then added to the flasks and placed on a shaker for 30 minutes. All experiments were conducted in triplicate. Concentration curves were determined with metal concentrations varying from 1.0 to 16 mg/L. The pH was buffered at pH 5 using 5 mM of sodium acetate after titrating to pH 5 with NaOH. The pH and temperature of the contact solution both before and after each experiment were measured. At the end of the experiment, the contents of the flasks were filtered (for vascular aquatic plants) or centrifuged (for microalgal samples) to separate the biomass from the solution. The biomass was then stored for later digestion and analysis. The filtrate or supernatant were immediately analyzed for metal concentration.

All samples were analyzed using a Perkin Elmer Model 3100 AAS. Before running the samples, a five-point standard curve was established. At the end of 20 samples, the 20th sample was rerun as a duplicate. A matrix spike was also performed before starting the next run of samples. The recoveries of the standards, duplicates, and matrix spike were within 85 to 115%.

Both dry weight and ash-free dry weight (AFDW) analyses were performed as per the Standard Methods for Water and Wastewater (AWWA 1993). The mean values from triplicate runs of both initial and final metal concentrations were used to calculate percent metal remaining in solution, percent metal removed by biomass, specific adsorption (mg metal/kg of dry weight biomass), and bioconcentration factor (specific adsorption/residual metal concentration).

RESULTS

A total of 16 biomass species were screened with initial metal concentration of Zn and Cd at 1 mg/L. The results show that some plants had significantly higher metal uptake than others. Figures 1 and 2 show the specific adsorption for Cd and Zn in these experiments. The most promising species were *Phormidium* (specific adsorptions of 1092 mg/kg Cd, and 823 mg/kg Zn) and water milfoil (specific adsorptions of 532 mg/kg Cd, and 722 mg/kg Zn); and they were used in further experiments with Ni, Pb, and Cu as well as Cd and Zn at various initial metal concentrations.

Adsorption isotherms were established for both plants. Adsorption characteristics were calculated using the Langmuir adsorption isotherm equation, $C/Y = C/Y_m + 1/kY_m$ where Y_m is the maximum specific adsorption, k the equilibrium constant, and Y the specific adsorption at residual metal concentration C. From

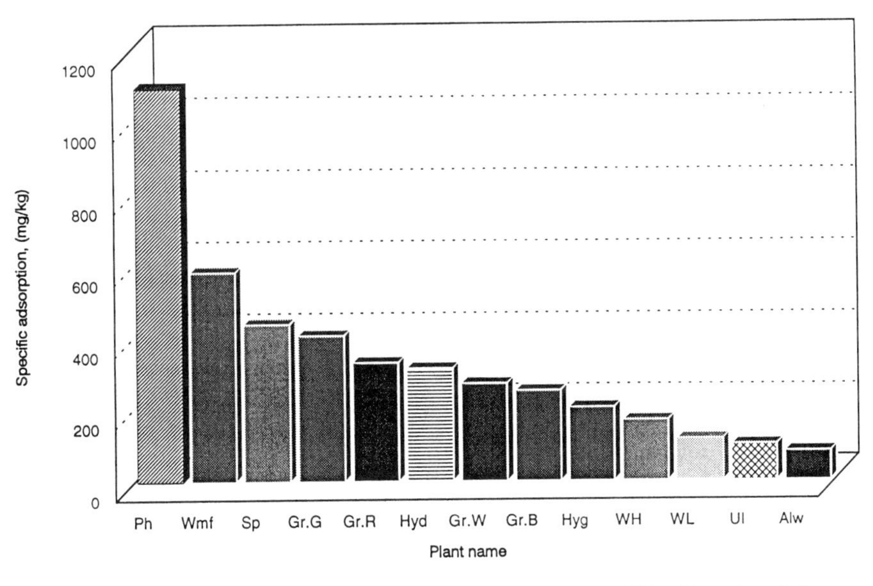

Ph - *Phormidium*
Wmf - Water milfoil
Sp - *Spirulina*
Gr.G - *Gracillaria* (green)
Gr.R - *Gracillaria* (red)
Hyd - *Hydrilla*
Gr.W - *Gracillaria* (wild)
Gr.B - *Gracillaria* (brown)
Hyg - *Hygrophyllum*
WH - Water hyacinth
WL - Water lettuce
Ul - Ulva
Alw - Alligator weed

FIGURE 1. Specific adsorption of cadmium onto various biomass. Initial Cd concentration = 1 mg/L. Biomass density was 0.01 kg/L wet wt for all biomass except for *Phormidium*, which was 94 mg/L dry wt.

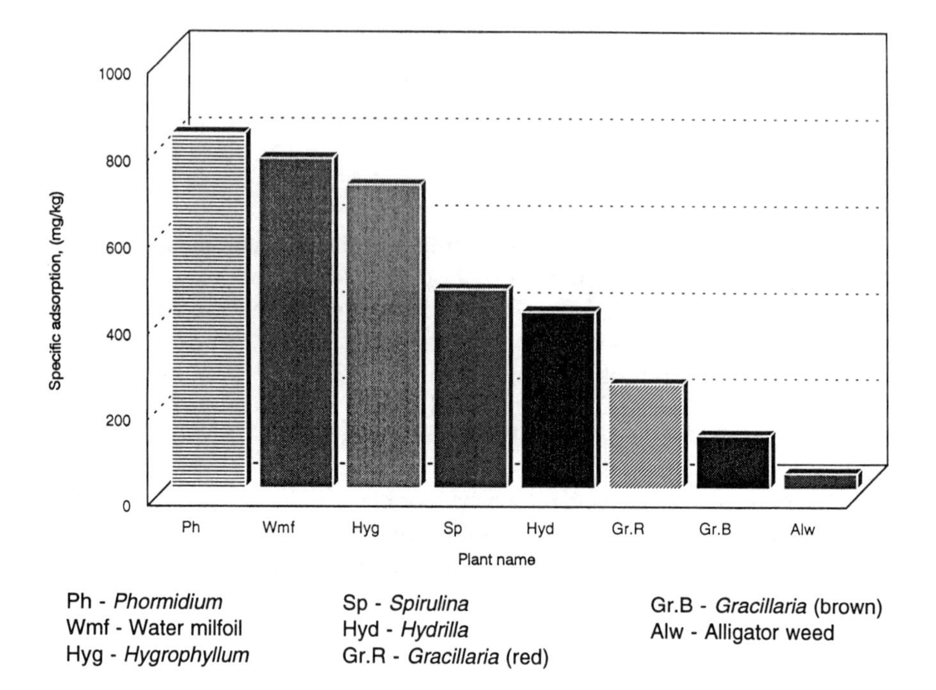

FIGURE 2. Specific adsorption of zinc onto various biomass. Initial Zn concentration = 1 mg/L. Biomass density was 0.02 kg/L wet wt for all biomass except *Phormidium*, 81 mg/L dry wt; *Hydrilla*, 0.1 kg/L; and alligator weed, 0.1 kg/L.

a plot of C/Y vs. C, the slope ($S = 1/Y_m$) gives Y_m and the intercept ($I = 1/kY_m$) gives k. Table 1 lists the maximum specific adsorption at pH 5 for five metals for each plant species. Water milfoil had the higher maximum specific adsorption of Pb, Zn, and Cu of 55,600, 13,500, and 12,900 mg/kg, respectively. Cd and Ni were adsorbed similarly by both plants. The lowest residual metal concentration actually measured are also shown in Table 1. The results indicate that the minimum residual concentration for the Cd, Zn, Ni, and Cu were about 0.01 mg/L except Pb which was below detection limit (0.004 mg/L).

CONCLUSION

Requirements for developing a practical bioremoval process include low-cost production of plant biomass, ease of removing the biomass from suspension, high maximum specific adsorption, and the capability to reduce metal concentration to very low residual values. Filamentous microalgae and macroscopic plants are easily harvested, and many are potentially productive in mass culture. The experimental data presented here indicate that these types of plants may also have the promising metal-adsorbing characteristics.

TABLE 1. Adsorption characteristics of aquatic plants at pH 5.

	Phormidium		Water Milfoil	
Metal	Maximum specific adsorption (mg/kg)	Minimum residual concentration (mg/L)	Maximum specific adsorption (mg/kg)	Minimum residual concentration (mg/L)
Cd	9,600±1,200	0.008±.001	8,200±1,000	0.007±0.001
Zn	9,400±1,100	0.010±0.004	13,500±3,400	0.100±0.004
Pb	13,600±1,900	<0.004	55,600±10,000	<0.004
Ni	5,700±600	0.011±0.002	5,800±800	0.029±0.002
Cu	10,100±800	0.009±0.007	12,900±2,500	0.008±0.007

REFERENCES

AWWA. 1992. *Standard Methods for Water and Wastewater.* AWWA, APHA, WPCF.

Chawla, G., J. Singh, and P. N. Viswanathan. 1991. "Effect of pH and Temperature on the Uptake of Cadmium by Lemna Minor." *Bul. Environ. Toxicol 47*: 84-90.

Darnall, D. W., B. Green, M. T. Henzl, J. M. Hosea, R. H. McPherson, J. Sneddon, and M. D. Alexander. 1986. "Recovery of Heavy Metals by Immobilized Algae." In R. Thompson (Ed.), *Trace Metal Removal from Aqueous Solution,* Whitstable Litho Ltd., pp. 1-24.

Feiler, H. D., and D. W. Darnall. 1991. *Remediation of Ground Water Containing Radionuclides and Heavy Metals Using Ion Exchange and the AlgaSORB Biosorbent System,* Final Report under Contract No. 02112413 DOE/CH-9212.

Green, B., and G. W. Bedell. 1990. "Immobilize Algae for Metal Recovery." In I. Akatsuka (Ed.), *Introduction to Applied Phycology.* Academic Publishing, The Hague, pp. 109-136.

Guilizzoni, P. 1991. "The Role of Heavy Metals and Toxic Materials in the Physiological Ecology of Submersed Macrophytes." *Aquatic Botany 41*:878-109.

Guven, K. C. 1992. "Metal Uptake by Black Sea Algae." *Botanica Marina 35*:337-340.

Muramoto, S., and Y. Ohi. 1983. "Removal of Some Heavy Metals from Polluted Water by Water Hyacinth." *Bul. Environmental. Contam. Toxicol. 30*:170-177.

Nakatsu, C., and T. C. Hutchinson. 1988. "Extreme Metal and Acid Tolerances of *Euglena mutabilis* and an Associated Yeast from Smoking Hills, Northwest Territories and Their Apparent Mutualism." *Microbial Ecology 16*:213-231.

Sela, M., A. Huttermann, and E. Tel-Or. 1990. "Studies on Cadmium Localization in the Water Fern Azolla." *Physiologia Plantarum 79*:547-553.

Stevens, S. E., Jr., K. Dionis, and L. R. Stark. 1989. "Manganese and Iron Encrustation on Green Algae in Acid Mine Drainage." In D. A. Hammer (Ed.), *Constructed Wetlands for Waste Water Treatment; Municipal, Industrial and Agricultural.* Lewis Publishers, Boca Raton, FL. pp. 765-773.

USEPA. 1986. *Quality Criteria for Aquatic Life.*

Whitton, B. A., M. G. Kelly, J.P.C. Harding, and P. J. Say. 1991. *Use of Plants to Monitor Heavy Metals in Freshwaters.* HMSO, London.

Wilde, E. W., and J. R. Benemann. 1993. "Bioremoval of Heavy Metals by the Use of Micro-algae." *Biotech. Adv. 11*:781-812.

Wolverton, B. C., and R. C. McDonald. 1975. *Water Hyacinth and Alligator Weeds for Removal of Lead and Mercury from Polluted Waters.* NASA Report.

Bioconversion of Cyanide and Acetonitrile by a Municipal-Sewage-Derived Anaerobic Consortium

Nick J. Nagle, Christopher J. Rivard,
Ali Mohagheghi, and George Philippidis

ABSTRACT

The production, use, and disposal of organonitriles, such as cyanide, acetonitrile, and acrylonitrile, represent a major environmental challenge. In this study, an anaerobic consortium was examined for its ability to adapt to and degrade the representative organonitriles, cyanide, and acetonitrile. Adaptation to cyanide and acetonitrile was achieved by adding increasing levels (50, 250, and 500 mg/L) of cyanide and acetonitrile to the anaerobic consortium, followed by extensive incubation over a 90-day period. The anaerobic consortium adapted most rapidly to the lower concentrations of each substrate and resulted in reductions of 85% and 83% of the cyanide and acetonitrile, respectively, at the 50 mg/L addition level. Increasing the concentration of both cyanide and acetonitrile resulted in reduced bioconversion. Two continuously stirred tank reactors (CSTR) were set up to examine the potential for continuous bioconversion of organonitriles. The anaerobic consortium was adapted to continuous infusion of acetonitrile at an initial concentration of 10 mg/L·day in phosphate buffer. The anaerobic consortium consistently demonstrated greater than a 79% reduction in acetonitrile added to the CSTR over the range of 25 to 150 mg/L·day. The performance of the acetonitrile-amended digester was similar to that of the control CSTR, to which phosphate buffer was added. Anaerobic treatment of nitrile wastes may be an attractive option. It results in significant destruction of wastes utilizing a simplified process design, containment of process gases, and potential revenue from the production of methane.

INTRODUCTION

In 1992, more than 4 million tons of organonitriles were produced (Basheer et al. 1992, Read et al. 1991, Read 1992). Organonitriles are used in plastics, ore leaching, coal processing, and metal plating. They are also used as intermediates in the production of rubber and pharmaceuticals (Mudder & Whitlock 1984).

TABLE 1. Industrial organonitrile production.

Nitrile	Formula	Total Production (10^6 U.S. pounds)	
		1990	1992
Hydrogen cyanide	HCN	572	1,244
Acetonitrile	C_2H_3N	17	42
Acrylonitrile	C_3H_3N	2,670	3,055
Other nitriles[a]	—	89	n/a

(a) Other nitriles include butyronitrile, propionitrile, adiponitrile, trichloronitrile, and benzonitrile.

Acrylonitrile accounts for over 50% of organonitriles produced (Table 1). The wastestreams that result from the manufacture of acrylonitrile are acetonitrile and cyanide. Acetonitrile represents 2% to 3% of the volume of acrylonitrile produced (Read 1992). Industrial use of cyanide results in the production of more than 3 billion L of cyanide-containing waste annually (Read et al. 1991).

The costs of recovering dilute acetonitrile wastes often are prohibitive. Therefore, more concentrated wastewaters may be incinerated to recover their BTU value. Dilute cyanide wastewaters can be treated by alkaline chlorination, or by oxidation using hydrogen peroxide. These treatments can be costly and can create toxic by-products that require posttreatment. Biological treatment of cyanide by aerobic microorganisms has been demonstrated successfully by several authors (Mudder & Whitlock 1984, 1983), but problems with accumulation of aerobic biomass in confined systems can reduce this degradation. Acetonitrile can also be easily degraded by aerobic microorganisms. Alternatively, there are few reports of successful anaerobic biological treatment of cyanide or acetonitrile in the literature (Fallon et al. 1991, Fedorak & Hyman 1987, Fallon 1992). Anaerobic treatment systems have yet to be demonstrated at any significant scale for cyanide and acetonitrile wastewaters. Anaerobic treatment of dilute nitrile wastewaters offers advantages in that the contained system does not require oxygen input, and process emissions are controlled through biogas by-product use in combustion systems. Thus, the anaerobic system controls potential volatile organic compounds (VOCs) while producing an energy by-product. In addition, anaerobic systems result in substantially lower levels of microbial biomass, reducing the need for nutrients.

Although the exact mechanism of anaerobic degradation of cyanide is not known, it is theorized that under anaerobic conditions cyanide is broken down to bicarbonate and ammonia, as shown in Equation 1.

$$HCN + 2H_2O \rightarrow HCOO^- + NH_4^+ \qquad \Delta G^\circ = -15.6 \text{ kcal/mol} \qquad (1)$$

Furthermore, formate produced during the anaerobic conversion of cyanide is metabolized by a variety of methanogenic bacteria to produce methane and carbon dioxide, as described in Equation 2 (Zeikus 1977):

$$4HCOO^- + 4H^+ \rightarrow CH_4 + 3CO_2 + 2H_2O \qquad \Delta G° = -34.7 \text{ kcal/mol} \qquad (2)$$

The summation of the two reactions results in the production of 1 mole of methane for every 4 moles of cyanide degraded.

Reactions similar to that for cyanide may be anticipated for the anaerobic bioconversion of acetonitrile to methane and carbon dioxide, as shown in Equations 3 and 4.

$$CH_3CN + 2H_2O \rightarrow CH_3COO^- + NH_4^+ \qquad \Delta G° = -17.9 \text{ kcal/mol} \qquad (3)$$

$$CH_3COO^- + H^+ \rightarrow CH_4 + CO_2 \qquad \Delta G° = -8.6 \text{ kcal/mol} \qquad (4)$$

For acetonitrile, anaerobic bioconversion should result in the production of 1 mole of methane for every mole of acetonitrile degraded.

MATERIALS AND METHODS

Chemical Reagents

Cyanide and acetonitrile used in anaerobic bioconversion studies were purchased from Sigma Chemical Co. (St. Louis, Missouri) and were reagent grade.

Biochemical Methane Potential (BMP) Analysis

The BMP assays were performed as previously described (Owens et al. 1979) to determine the ultimate methane yields from the anaerobic conversion of the feedstocks by the anaerobic consortium. Studies were conducted in 155-mL serum bottles and mixed on an orbital shaker at 37°C. Active anaerobic inoculum for the BMP assays was obtained from low-solids anaerobic digesters at the Denver Municipal Waste Water Reclamation Plant in Denver, Colorado. The effluent from these low-solids digesters was screened through a 1.0-mm United States Standard Testing Sieve. The serum bottles were then inoculated with 100 mL of this screened effluent. The anaerobic bioconversion of potassium cyanide and acetonitrile was evaluated by adding 50, 250, and 500 mg/L of each compound (separately) into BMP test bottles. All experimental tests were performed in triplicate. Biogas production was measured using a pressure transducer fitted to a 22-gauge needle for penetration into serum bottles. After each pressure measurement cycle, the remaining overpressure was released from the serum bottles. BMP assays were incubated for 90 days to ensure ultimate biodegradation. Controls for the experiment included serum bottles that contained autoclaved inoculum (anaerobic sewage sludge), supplemented with 500 mg/L of either filtered sterilized acetonitrile or cyanide. The controls were identical to the experimental serum bottles except the biological activity had been destroyed by autoclaving.

Low-Solids Digester Operation

Two low-solids anaerobic digesters were operated over a period of 5 months, as described previously (Rivard 1993). Each digester has a 3.5-L working volume and semicontinuously stirring (15 min of each 0.5 h). The digesters were maintained at 37°C in a constant temperature warm room. The anaerobic digesters were batch-fed daily by adding the relatively dry municipal solid waste (MSW) feedstock and a liquid nutrient solution. In the batch feeding protocol, a volume of feed was removed daily to maintain the reactor sludge volume of 3.5 L. In the operation of the reactors, the solids retention time was equivalent to the hydraulic retention time. A syringe pump (Harvard Apparatus Co., South Natick, Massachusetts, model #2265), was used to continuously amend the digesters with either acetonitrile in buffer or buffer alone over a 24-h period.

Analysis of Chemical Oxygen Demand

Chemical oxygen demand (COD) for acetonitrile was determined as previously described (APWA et al. 1980). The COD assay employed the microdetermination method using commercially available "twist tube" assay vials (Hach, Loveland, Colorado).

Cyanide and Acetonitrile Analysis

The analysis of cyanide refers to "total cyanide" and includes cyanide complexed to organics and metals, as well as free cyanide. Samples were first treated to remove sulfide, distilled, then analyzed for cyanide by chlorination. Nonamended controls as well as spiked samples were included in the analysis to assure quality control. Levels of acetonitrile were determined using gas-liquid chromatography (GLC). A Hewlett-Packard Model 5840A gas chromatograph equipped with a flame ionization detector, a Model 7672A autosampler, and a Model 5840A integrator (Hewlett-Packard) was used. The chromatograph was equipped with a glass column packed with Supelco 60/80, Carbopack C/0.3%, and Carbowax 20M/0.1% H_3PO_4 for efficient separation.

Gas Analysis

The composition of methane and carbon dioxide in the biogas produced was determined using gas chromatography. For this analysis, a Gow-Mac (Model 550) gas chromatograph equipped with a Porapak Q column and a thermal conductivity detector with an integrating recorder was used.

Theoretical Methane Yield

The theoretical methane yield for the added substrates, cyanide and acetonitrile, was calculated from COD values as previously described (Rivard 1993). The ratio of actual methane yields for a given anaerobic fermentation to the theoretical methane yield calculated from COD values is a direct reflection of the organic carbon conversion of the substrate (i.e., cyanide and acetonitrile) added.

RESULTS

Adaptation Studies

To study the effects of shock loadings on the rate of adaptation and anaerobic bioconversion of test substrates, a mixed culture of anaerobic bacteria was obtained from the digestion of municipal sewage. Cyanide and acetonitrile were added to the anaerobic consortium at 50, 250, and 500 mg/L. The addition of cyanide, at all concentrations, resulted in an immediate inhibition of gas production as compared to the controls (see Figure 1). However, following a 10-day period of adaptation, gas production returned to similar, but still lower, levels than the controls. Further adaptation resulted in biogas production rates that exceeded control levels for the 50 and 250 mg/L cyanide addition rates. Although the biogas production levels never exceeded control levels in the 500 mg/L cyanide amended cultures, active biogas production was measured throughout the 90-day incubation period.

Data for anaerobic bioconversions amended with acetonitrile are shown in Figure 2. The data indicate that except for an initial burst of biogas production immediately following acetonitrile addition, all amendments demonstrated transient inhibition followed by active biogas production similar to that determined for cyanide addition. However, the effects on biogas production resulting from acetonitrile addition were less pronounced than those observed with cyanide

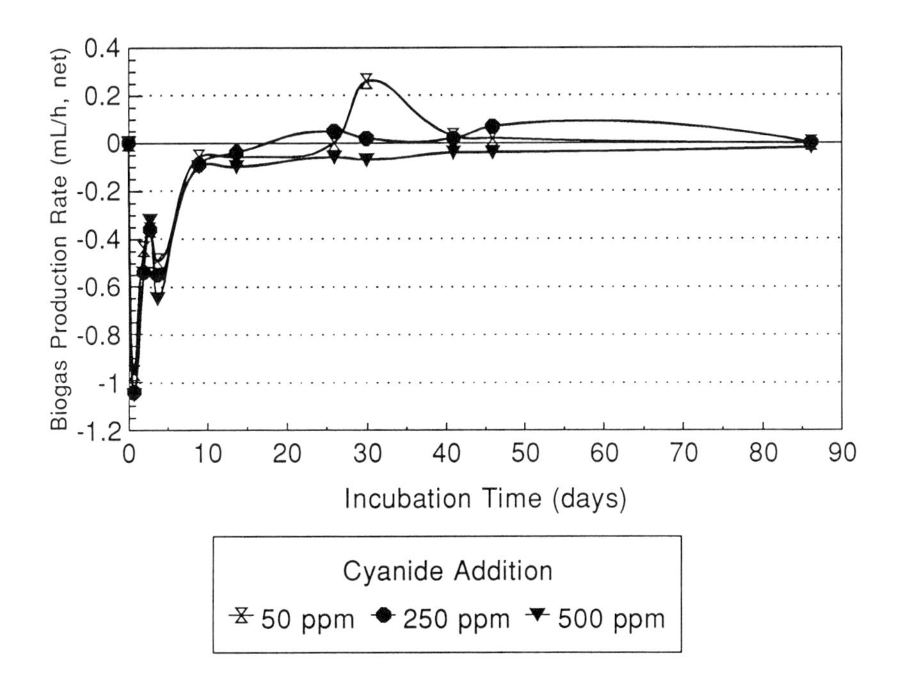

FIGURE 1. Biochemical methane potential (BMP) assay cyanide conversion.

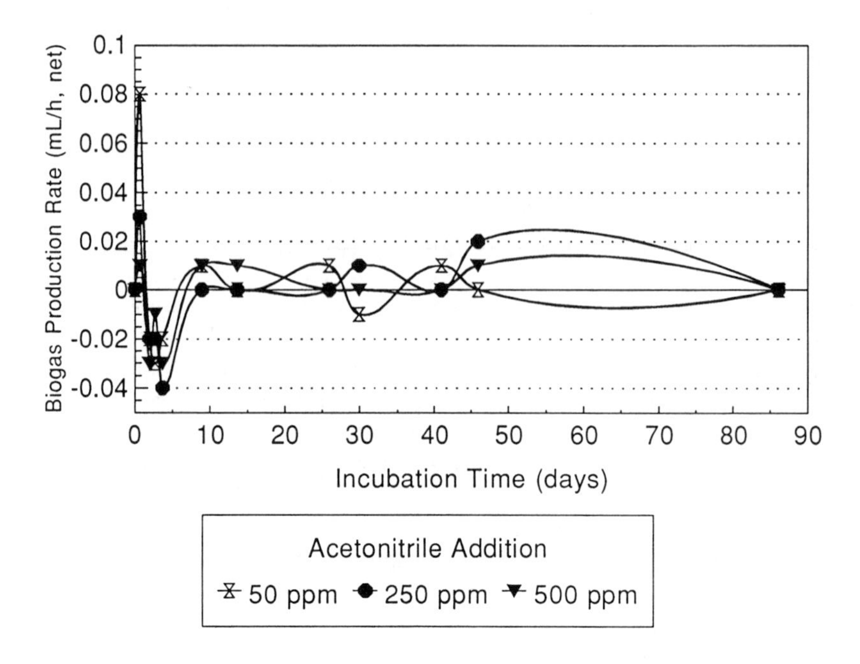

FIGURE 2. Biochemical methane potential (BMP) assay acetonitrile conversion.

addition. Additionally, biogas production exceeded control levels in samples amended with 500 mg/L acetonitrile.

Table 2 summarizes the removal of added cyanide and acetonitrile from BMP assays using both anaerobic biodegradation estimates from biogas yields and direct chemical analysis. The data indicate that cyanide does not appear

TABLE 2. Summary of anaerobic bioconversion estimates from BMP biogas production and direct chemical analysis.

Addition		% Degradation (methane yield)	% Degradation (Chemical Analysis)
Cyanide	50 mg/L	0	85.0 ± 8.8
	250 mg/L	0	44.8 (1 sample only)
	500 mg/L	0	0
(control)	500 mg/L	0	0
Acetonitrile	50 mg/L	150.6	83.4 ± 1.8
	250 mg/L	40.0	6.4 ± 2.2
	500 mg/L	17.1	4.4 ± 4.2
(control)	500 mg/L	0	0

to be converted in the BMP assay based solely on the biogas production yields. However, direct chemical analysis for cyanide indicates that for the 50 mg/L addition level, most of the cyanide was biologically converted. In addition, for at least one of the 250 mg/L BMP replicates, greater than 44% of the cyanide was converted. The loss of cyanide from the liquid phase in BMP assays was determined to reflect direct microbial conversion rather than simple adsorption, as indicated by the quantitative recovery of cyanide from the controls.

The anaerobic bioconversion of acetonitrile reflected complete conversion based on biogas yields for the 50 mg/L addition level. In fact, the biogas yield appeared to overestimate the level of conversion for acetonitrile at the 50 mg/L addition level. Substantial biological conversion was also demonstrated for the higher levels of acetonitrile addition, although increasing concentrations appear to reduce biogas yields in this initial assay. The data for acetonitrile conversion, as determined by direct chemical analysis, indicate that although the majority of the acetonitrile was converted in the 50 mg/L addition samples, insignificant conversion of acetonitrile occurred at the 250 mg/L and 500 mg/L addition levels. As in the cyanide experiment, little if any acetonitrile was removed by absorption, as demonstrated by the controls.

Continuous Addition of Acetonitrile to an Anaerobic CSTR System

The continuous anaerobic bioconversion of acetonitrile was examined using conventional low-solids CSTR systems fed a processed MSW feedstock. The digesters were supplemented with MSW to provide nutrients and carbon for the growth and maintenance of the microorganisms. MSW is a complex feedstock that provides supplemental carbon and nitrogen that they could not receive from the low levels of acetonitrile addition or phosphate addition alone. One digester system received acetonitrile in phosphate buffer; the control system received only phosphate buffer addition. Substrates were administered continuously using a syringe pump. Both CSTR systems demonstrated comparable fermentation parameters of sludge pH, biogas methane content, volatile fatty acid pools, MSW feedstock conversion, and free ammonia levels. The average conversion of acetonitrile in the amended CSTR as determined by chemical analysis is listed in Table 3. The data indicate an average conversion of 79.4% ± 11.1 over the amendment range of 10 to 150 mg/L·day. Incremental increases in the amendment rate of acetonitrile did not result in inhibition of the fermentation system at any time during the experimentation. Instead, increasing the acetonitrile amendment rate to the CSTR system initially resulted in reduced conversion levels, followed by rapid adaptation with increasing conversion yields.

DISCUSSION

Few reports exist on the anaerobic bioconversion of cyanide. Even less is known for acetonitrile. One study suggests that the anaerobic bioconversion of

TABLE 3. Anaerobic bioconversion of added acetonitrile in a CSTR system.

Addition Rate (mg/L·d)	Days	% Bioconversion
10	20	76.4 ± 8.7
25	7	70.9 ± 13.0
50	7	65.3 ± 14.9
100	60	82.2 ± 12.7
150	40	76.3 ± 7.2

cyanide is indeed possible but may be limited by its soluble concentration above 5 mg/L (Dilek & Yetis 1992). Furthermore, Fallon and his colleagues demonstrated the anaerobic bioconversion of cyanide at 300 mg/L in a digester system operated with a hydraulic retention time of 25 days (Fallon et al. 1991). In this study, we determined that shock loadings of cyanide and acetonitrile in the range of 50 to 500 mg/L resulted initially in a transient inhibition of biogas production (approximately 10 days). Following adaptation, gas production returned to or exceeded control levels. Using the BMP assay, the conversion of the nitrile was either underestimated (i.e., cyanide) or overestimated (i.e., acetonitrile) by determination of biogas production. This may be somewhat expected by the low level of test substrate addition relative to the background biogas production in the assay system. A more accurate determination of nitrile bioconversion results from the chemical analysis of effluent samples from assay and CSTR tests.

In continuous infusion experiments, anaerobic CSTR systems rapidly adapted to the addition of acetonitrile at the 10 mg/L·day level. Increasing the rate of acetonitrile addition resulted in comparable conversion rates following adaptation. Although the CSTR system is currently operated at an acetonitrile amendment rate of 150 mg/L·d, we will pursue increasing the amendment rate, attempting to achieve high rates of bioconversion at rates of 300 to 500 mg/L·d (comparable to that achieved by Fallon et al. 1991, for cyanide). Increasing the level of conversion of acetonitrile may be further improved through additional adaptation of the anaerobic consortium.

CONCLUSION

Anaerobic treatment systems may represent an attractive option to conventional treatments. Anaerobic systems can offer high conversions of industrial wastes, a simplified process design, and containment of volatile gases. Reduced treatment costs may be achieved through reductions in energy needed for operation and potential revenues from energy derived from the production of methane. Deploying anaerobic digestion systems such as fixed-film, upflow anaerobic sludge blanket reactor (UASBR) and sequencing batch reactor (SBR) systems, which separate the hydraulic detention time to allow for a higher retention of the microbial catalyst, may also improve the conversion level. If this technology is to be applied to industrial wastestreams, demonstration of effective anaerobic bioconversion of acetonitrile wastewaters in the concentration range of 100 to 500 mg/L is required. Further economic evaluation of the anaerobic system will serve to justify its application over competing disposal methods.

REFERENCES

APWA-AWWA-WPCF. 1980. "Oxygen Demand (Chemical)." In *Standard Methods for the Examination of Water and Wastewater, 15th ed.* pp. 489-493. APHA, Washington, DC.

Basheer, O. M., J. E. Prenosil, and J. R. Bourne. 1992. *Biotechnology and Bioengineering 39*: 629-634.

Dilek, F. B. and U. Yetis. 1992. *Water Science Technology, 26*: 801-813.

Fallon, P. M. 1992. *Applied and Environmental Microbiology, 58*: 3163-3164.

Fallon, R. D., D. A. Copper, R. Speece, and M. Henson. 1991. *Applied and Environmental Microbiology, 57*: 1656-1662.

Fedorak, P. M., and S. E. Hyman. 1987. *Water Resources, 19*: 67-76.

Johnson, W. K., Schwendener, H., and Y. Yoshida. 1992. "Hydrogen Cyanide 664.5000h." *Chem. Econ. Hand.* SRI Publication, Menlo Park, CA.

Mudder, T. L., and J. L. Whitlock. 1983. In *Proceedings of the 38th Annual Purdue Industrial Waste Conference*, pp. 279-287.

Mudder, T. L., and J. L. Whitlock. 1984. U.S. Patent 4,461834.

Owens, W. F., D. C. Stuckey, J. B. Healy, L. Y. Young, and P. L. McCarty. 1979. *Water Res. 13*: 485-492.

Read, C. S. 1992. "Acrylonitrile 607.5000a." *Chem. Econ. Hand.* SRI publication, Menlo Park, CA.

Read, C. S., J. Riepl, and Y. Sakuma. 1991. "Acetonitrile 605.5000a." *Chem. Econ. Hand.* SRI Publication, Menlo Park, CA.

Rivard, C. J. 1993. *Appl. Biochem. and Biotech. 39/40*: 71-82.

United States Inter. Trade Comm. 1990. *Synthetic Organic Chemicals, U.S. Prod. and Sales.* USTIC Pub. 2470, Washington, DC.

Zeikus, J. G. 1977. *Bacteriological Reviews, 41*: 514-541.

Effect of Chemical Pretreatment on the Biodegradation of Cyanides

Boris N. Aronstein, James R. Paterek,
Laura E. Rice, and Vipul J. Srivastava

ABSTRACT

The application of Fenton's reagent (H_2O_2; Fe^{2+}) as a chemical pretreatment for acceleration of biological degradation of cyanides in soil-containing systems has been studied. In slurries of topsoil freshly amended with radiolabeled free cyanide ($K^{14}CN$) at pH 7.2, about 100% of the compound was removed from the system by the combination of chemical oxidation and biodegradation. In slurry of manufactured gas plant (MGP) soil, the extent of combined chemical-biological treatment was 50%. At the same time, approximately 15% of the cyanide was lost from the system by protonation and evolution of formed HCN. In slurries of both topsoil and MGP soil amended with radiolabeled $K_4[Fe(CN)_6]$, less than 20% was degraded. In soils previously equilibrated with free and complex cyanide, the highest extent of degradation resulted from chemical-biological treatment did not exceed 15%. To avoid massive evolution of HCN, the cyanide-amended topsoil was maintained at a pH of 10.0. At this pH, nearly 35% of the cyanides were removed from the system by combined chemical-biological treatment.

INTRODUCTION

Cyanides belong to a group of pollutants with a high level of metabolic toxicity (Knowles 1976). They are present mainly in contaminated soils and wastewaters in the form of complexes with transition metals or iron-cyanide minerals, such as "Prussian blue" (Meeussen et al. 1992a, Theis et al. 1994). Most of these complexes are chemically inert and thus much less toxic than the free form. However, in the range of pHs and redox potentials that may be found in the environment, the complex cyanides can slowly decompose to the toxic free cyanide ions. Recently, positive results have been obtained on the ability of microbial strains to degrade metal-cyano complexes (Finnegan et al. 1991, Silva-Avalos et al. 1990). It is recognized, however, that instead of direct decomposition of the complex forms of cyanide, microorganisms might utilize the free form produced by the decomposition of the metal-cyano complexes at a favorable pH and redox potential (Aronstein et al. 1994a, Meeussen et al. 1992b). To overcome the natural recalcitrance of cyanide species in industrial solid wastes and wastewaters and to aid remediation, the application of Fenton's reagent has been proposed. It was found that Fenton's reagent was able to produce hydroxyl radicals at a wide range of pHs, resulting in substantial removal of cyanides from aqueous and soil-containing systems (Aronstein et al. 1994b). The current study evaluates the effect of chemical pretreatment on the effectiveness of biodegradation at neutral and alkaline pHs. The results of this study provide evidence that application of Fenton's reagent for the partial oxidation of cyanides may be an important prerequisite for subsequent aerobic microbial degradation of these pollutants.

EXPERIMENTAL PROCEDURES AND MATERIALS

All chemicals were reagent-grade. Potassium cyanide, potassium ferrocyanide and ferrous sulfate were purchased from Aldrich Chemical Co. (Milwaukee, Wisconsin); $K^{14}CN$ (13.1 mCi/mmol, purity >98%) and 3-cyclohexylamino-1-propanesulfonic acid (CAPS) were purchased from Sigma Chemical Co. (St. Louis, Missouri). The solution of radiolabeled potassium ferrocyanide $K_4[Fe(C^{14}N)_6]$ was prepared as described previously (Aronstein et al. 1994a).

Aerobic microorganisms capable of using cyanide as a sole nitrogen source were isolated from cyanide-contaminated MGP soil (Aronstein et al. 1994a). Ten grams of soil were added to 100 mL of the salts medium (pH 7.2) and incubated at 22°C. After several weeks of incubation, the supernatant was decanted, centrifuged to collect cells, and used to inoculate salts medium containing glucose (1%) and KCN (25 mg/L) as the sole nitrogen source. To acclimate this consortium to higher pHs, an inorganic salts medium (ISM) prepared with 0.02M CAPS buffer was used. In microbial degradation studies, the consortium was pregrown for 3 to 5 days in ISM with potassium cyanide (25 mg/L) as the nitrogen source and glucose (1%) as the carbon source, on a rotary shaker operating at approximately 120 rpm. The medium was buffered by phosphate solution (to obtain pH 7.2) or by CAPS (for pH 10.0). The cells were harvested by centrifugation (10,000 rpm for 10 min at 4°C), the pellet was washed three times to remove residual cyanide, and the cells were resuspended in fresh salts medium. The culture used in the biodegradation experiments was incubated at 30°C on a rotary shaker operating at 140 rpm.

For chemical-biological degradation experiments using Fenton's reagent, 160-mL serum bottles were modified and the measurements of $H^{14}CN$ and $^{14}CO_2$ were conducted as described previously (Aronstein et al. 1994b). For pH 7.2 and pH 10.0, phosphate buffer and CAPS buffer were used respectively. Optimal concentrations of Fenton's constituents were determined previously and were used in all experiments—5% H_2O_2 (w/w) and 0.1 mM $FeSO_4$ (Aronstein et al. 1994b) were used in all experiments. To prevent microbial growth in the control, soils were sterilized by irradiation with ^{60}Co (1.4 Mrad). The equilibration of soils with radiolabeled chemicals to study the effect of weathering on degradation was described previously (Aronstein et al. 1991). To ensure the complete release of dissolved $^{14}CO_2$, at each time point, two bottles were sacrificed and the solutions were acidified to pH 4.0. Cultures were added to the experimental systems 120 hours after the initiation of chemical treatment. It was determined that after this period of time H_2O_2 was depleted to trace amounts under the conditions of the experiments.

RESULTS

The effect of soil particles on the rate and extent of cyanide removal from the system was determined in the presence of 10% slurries of MGP and topsoil at neutral pH. In 234 hours, approximately 24 to 28% of free and 5 to 8% of complex cyanide was removed from both soil slurries in the presence of aerobic mixed culture (AMC) (Table 1). The differences observed between the samples with or without cells were not statistically significant. Application of Fenton's reagent to topsoil resulted in the mineralization of 77% of $K^{14}CN$, while for both topsoil and MGP soil 14% of radioactivity added in the form of $K_4[Fe(^{14}CN)_6]$ was recovered as $^{14}CO_2$. (From here on, the evolution of $H^{14}CN$ from the control bottles is taken into account.) The application of AMC after Fenton's reagent resulted in removal of 95% of the free and 17% of complex cyanide from the topsoil. In MGP soil, sequential chemical-biological treatment resulted in the removal of about 50% of the free cyanide and 6% of the complex cyanide.

The microbiological removal of cyanides in 10% slurries containing topsoil and MGP soil previously exposed to the free and complex form, was much less than observed in the soil systems freshly amended with cyanide (Table 1). In the presence of AMC, only 11.0% of free and 1.1% of complex cyanide was removed from the topsoil slurry in 336 hours. The removal of cyanides from the MGP soil slurries previously exposed to the free and complex form of cyanide was even less than that of the topsoil slurries. Biological removal of cyanides from MGP soil slurries previously exposed to the free and complex forms of cyanide did not occur. Also, chemical treatment had little or no effect on cyanide removal. Sequential chemical-biological treatment, on the other hand, resulted in the removal of approximately 14.1% and 6.1% of the free and complex forms of cyanide, respectively.

To avoid massive evolution of HCN, a high pH (10.0) was applied to the slurries of topsoil that were freshly amended with cyanides (Figure 1). The overall extent of mineralization for free and complex cyanides subjected to the sequential chemical-biological treatment was about 35%, which is approximately eight times greater compared to the biological treatment alone. The degradation curves for both cyanides tested were similar, showing biphasic patterns. Chemical pretreatment also resulted in greater initial degradation rates after inoculation of the experimental system with cyanide-degrading microorganisms.

TABLE 1. Removal of [14-C] cyanides from soil slurries, measured by evolution of $H^{14}CN$ and $^{14}CO_2$ (% of initially added [14-C] cyanide).

	Soils Freshly Amended with Cyanides				Soils Preequilibrated with Cyanides			
	Topsoil		MGP Soil		Topsoil		MGP Soil	
Treatment[(a)]	KCN	$K_4[Fe(CN)_6]$	KCN	$K_4[Fe(CN)_6]$	KCN	$K_4[Fe(CN)_6]$	KCN	$K_4[Fe(CN)_6]$
Biological (234 h)								
Control[(b)]	18.0±2.1	1.9±0.3	28.6±1.5	10.0±0.9	2.7±0.1	1.4±0.2	3.7±1.1	1.1±0.2
Live cells[(c)]	23.7±3.7	4.7±0.5	27.8±1.5	8.3±1.3	13.7±1.5	2.5±0.3	3.5±0.3	1.5±0.3
Chemical (120 h)								
Control[(b)]	6.5±0.6	1.9±0.2	24.0±2.0	8.2±0.5	1.3±0.2	0.9±0.3	2.7±0.9	0.8±0.1
Fenton's[(c)]	83.4±2.8	13.7±0.8	76.7±4.4	14.5±0.8	9.0±1.0	1.8±0.5	4.0±0.7	1.1±0.1
Sequential Chemical-Biological (354 h)								
Control[(b)]	16.0±1.1	2.4±0.4	34.3±2.9	8.9±0.2	9.3±0.7	1.9±0.7	1.9±0.4	1.7±0.3
Cells after Fenton's[(c)]	100.0±5.5	19.2±1.3	84.3±1.7	15.3±1.1	13.4±0.4	2.7±0.5	15.9±2.1	7.8±0.7

(a) pH of the experimental systems 7.2.
(b) Values for control bottles (sterile soil — no treatment) represent the extent of protonation (measured by $H^{14}CN$ evolution).
(c) Values for treated bottles represent combined extent of protonation and degradation (measured by combined ^{14}HCN and $^{14}CO_2$ evolution).

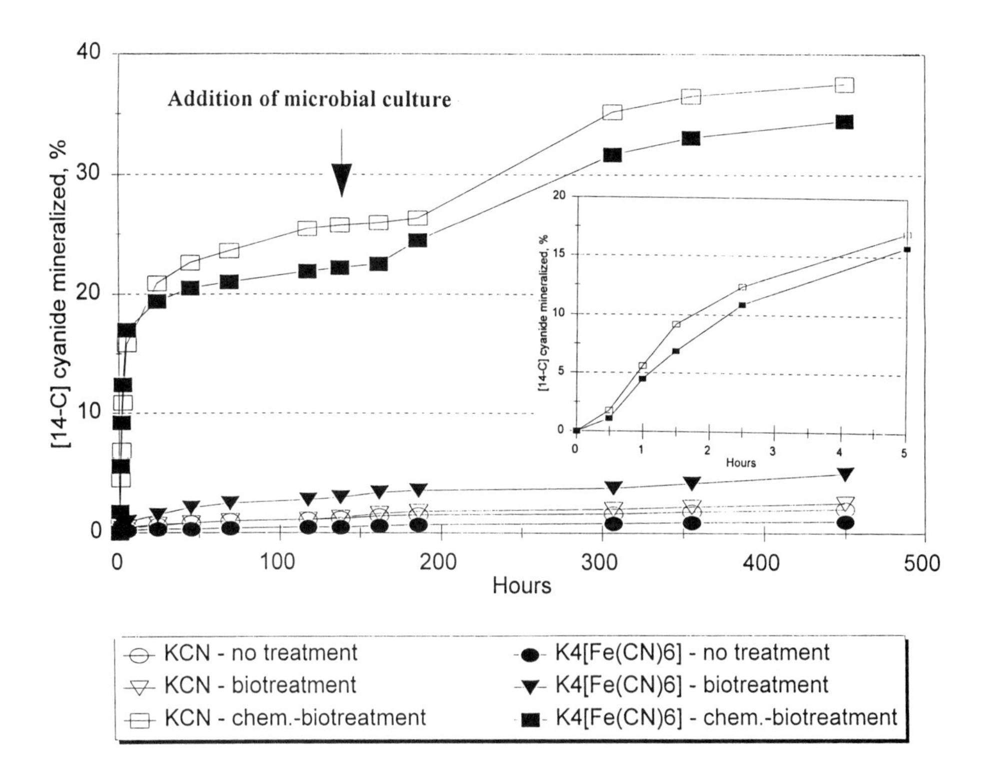

FIGURE 1. Removal of [14-C] cyanides from the freshly amended slurry of topsoil by sequential chemical-biological treatment at alkaline pH.

DISCUSSION

Although the data on the biodegradation of free cyanides resulting in the production of NH_3 and CO_2 at neutral pH is abundant in the literature (Raef et al. 1977, White et al. 1988), relatively little attention has been paid to the fact that, at pHs below 9.36 (pKa of HCN), the protonation of cyanide ions and consequent evolution of toxic gaseous hydrogen cyanide is taking place. The present study demonstrates limited ability of a microbial consortium acclimated to alkaline pHs, to degrade free cyanide without concurrent substantial evolution of HCN. The consortium developed was also able to degrade free and complex cyanides altered by chemical oxidation at alkaline pH. At the same time, no degradation of complex cyanide by microorganisms was observed without chemical pretreatment. This still leaves the question of the direct microbial attack on metal-cyano complexes open. Previously obtained data on biological degradation of complex cyanides (Finnegan et al. 1991, Silva-Avalos et al. 1990) were based on the assumption that no free form was present in the experimental systems amended initially with the chemically pure complex form. However, the specific redox and pH values of the experimental systems were not taken into account. It is known that these values are responsible for the distribution of the different cyanide species, including its free form. In a study on the chemical stability and decomposition rate of iron cyanide complexes in soil solutions, it was suggested that the degradation of the metal-cyano complexes may be due to chemical decomposition under the particular redox potential of the experimental systems rather than direct microbial degradation (Meeussen et al. 1992a). It was also shown that even relatively low amounts of free cyanide produced by chemical decomposition may support microbial growth of cyanide degraders in the systems amended initially with the complex form (Knowles & Bunch 1986).

The data obtained in a study with Fenton's reagent applied to the systems containing complex cyanide showed that the hydroxyl radicals were unable to oxidize the $Fe(CN)_6^{4-}$ ions to ammonia and carbon dioxide (Aronstein et al. 1994b). In soils, in addition to complexation by the soil's heavy metals, cyanides are directly sorbed by the solid surfaces and are therefore less available for chemical conversion and biological degradation with time (Aronstein et al. 1994a). At the same time, partial oxidation of potassium hexacyanoferrate (II) might result in the formation of more weak metal-cyano complexes or other less recalcitrant compounds. Previous direct observations in oxidation of ferrocyanide by hydrogen peroxide established the formation of the stable product $Fe(CN)_6^{3-}$ (Buxton et al. 1988). This ion might be the subject of further dissociation with the production of free cyanide (Cherryholmes et al. 1985).

REFERENCES

Aronstein, B. N., Y. M. Calvillo, and M. Alexander. 1991. "Effect of Surfactants at Low Concentrations on the Desorption and Biodegradation of Sorbed Aromatic Compounds in Soil." *Environmental Science and Technology 25*: 1728-1731.

Aronstein, B. N., A. Maka, and V. J. Srivastava. 1994a. "Chemical and Biological Removal of Cyanides from the Aqueous and Soil-Containing Systems." *Applied Microbiology and Biotechnology 40*: 700-707.

Aronstein, B. N., R. A. Lawal, and A. Maka. 1994b. "Degradation of Cyanides by Fenton's Reagent in Aqueous and Soil-Containing Systems." *Environmental Toxicology and Chemistry 13*: 1719-1726.

Buxton, G. V., C. L. Greenstock, W. P. Helman, and A. B. Ross. 1988. "Critical Review of Rate Constants for Reactions of Hydrated Electrons, Hydrogen Atoms and Hydroxyl Radicals ($\cdot OH/\cdot O^-$) in Aqueous Solutions." *Journal of Physical and Chemical Reference Data 17*: 513-886.

Cherryholmes, K. L., W. J. Cornils, D. B. McDonald, and R. C. Splinter. 1985. "Biological Degradation of Iron Cyanides in Natural Aquatic Systems." In R. D. Cardwell, R. Purdy, and R. C. Bahner (Eds.), *Aquatic Toxicology and Hazard Assessment, Seventh Symposium ASTM STP854 502-511*, pp. 502-511. American Society for Testing Materials, Philadelphia, PA.

Finnegan, I., S. Toerien, L. Abbot, F. Smith, and H. G. Raubenheimer. 1991. "Identification and Characterization of an *Acinetobacter* sp. Capable of Assimilation of Cyano-Metal Complexes, Free Cyanide Ions and Organic Nitriles." *Applied Microbiology and Biotechnology 36*: 142-144.

Knowles, C. J. 1976. "Microorganisms and Cyanide." *Bacteriological Reviews 40*: 652-680.

Knowles, C. J., and A. W. Bunch. 1986. "Microbial Cyanide Metabolism." *Advances in Microbial Physiology 27*: 74-111.

Meeussen, J.C.L., G. K. Meindert, W. H. van Reimsdijk, and F.A.M. de Haan. 1992a. "Dissolution Behavior of Iron Cyanide (Prussian Blue) in Contaminated Soils." *Environmental Science and Technology 26*: 1832-1838.

Meeussen, J.C.L., M. G. Keizer, and F.A.M. de Haan. 1992b. "Chemical Stability and Decomposition Rate of Iron Cyanide Complexes in Soil Solutions." *Environmental Science and Technology 26*: 511-516.

Raef, S. F., M.A.K. Characklis, and C. H. Ward. 1977. "Fate of Cyanide and Related Compounds in Aerobic Microbial Systems. II. Microbial Degradation." *Water Resources 11*: 485-492.

Silva-Avalos, J., M. G. Richmond, O. Nagappan, and D. A. Kunz. 1990. "Degradation of the Metal-Cyano Complex Tetracyanonickelate (II) by Cyanide-Utilizing Bacterial Isolates. *Applied and Environmental Microbiology 56*: 3664-3670.

Theis, T. L., T. C. Young, M. Huang, and K. C. Knutsen. 1994. "Leachate Characteristics and Composition of Cyanide-Bearing Wastes From Manufactured Gas Plants." *Environmental Science and Technology 28*: 99-106.

White, J. M., D. D. Jones, D. Huang, and J. J. Gauthier. 1988. "Conversion of Cyanide to Formate and Ammonia by a Pseudomonad Obtained From Industrial Wastewater." *Journal of Industrial Microbiology 3*: 263-372.

Biological Reduction of Soluble Selenium in Subsurface Agricultural Drainage Water

Lawrence P. Owens, Kurt C. Kovac,
Jo Anne L. Kipps, and Desmond W. J. Hayes

ABSTRACT

Over 100,000 acre-feet of agricultural drainage water is generated annually in the San Joaquin Valley, California. This brackish water contains elevated concentrations of nitrate and selenium. The selenium in the agricultural drainage water is in the soluble, selenate (+6 oxidation state) form. Selenium has been shown to be detrimental to wildlife. A multistaged biological selenium removal process consisting of an upflow anaerobic sludge blanket reactor (UASBR), followed by a fluidized-bed reactor and slow-sand filter, has been researched. Selenium removal is achieved by biological reduction to selenite (+4 oxidation state), then to particulate elemental selenium, which accumulates in reactor sludge or exits with reactor effluent. Denitrification and selenium removal coincide. This paper addresses two operational phases of the first stage UASBR. Phase I focused on adapting imported granular sludge for selenium removal. Phase II began with in situ development of UASBR granular sludge, followed by long-term reactor operation. The UASBR has achieved soluble selenium-removal rates of greater than 90 and 80% for Phase I and II operations, respectively.

INTRODUCTION

Over 100,000 acre-feet of agricultural drainage water is generated annually in the San Joaquin Valley, California. This brackish water contains elevated concentrations of nitrate and selenium (in excess of 40 mg/L and 500 µg/L, respectively). The selenium in the agricultural drainage water is in the soluble, selenate (+6 oxidation state) form. Selenium has been shown to be detrimental to wildlife and increasing salt accumulations threaten the productivity of more than 4,000 km^2 of irrigated farmland. Past investigations by EPOC Ag removed selenium from agricultural drainage water through a biological staged-reactor configuration, beginning with a UASBR, followed by a fluidized-bed reactor, then with a standard solids separation process. In the process, selenate is biologically reduced to selenite (+4 oxidation state) and further to elemental selenium. Denitrification and selenium reduction coincide.

Initial work on this process by EPOC Ag was short-lived and was characterized by fluctuating operating conditions but was determined to show the most promise for further investigation and development among the competing alternatives. Evaluation of these biological processes for feasibility requires long-term pilot testing at steady operating conditions, which is currently being performed at the Adams Avenue Agricultural Drainage Research Center.

ADAMS AVENUE BIOLOGICAL SELENIUM-REMOVAL TEST PROGRAM

The California Department of Water Resources is sponsoring the Engineering Research Institute at California State University, Fresno, to perform research on pilot-scale, biological selenium-removal processes at the Adams Avenue Agricultural Drainage Research Center located about 70 km west of Fresno, California. Additional assistance is provided by the U.S. Bureau of Reclamation and Westlands Water District. The first phase of testing used an 11-m^3 conical-bottom UASBR with an internal gas-solid-liquid separator and an "imported" granular sludge for startup (Kipps 1994). In the second phase of work, additional reactors were brought on line in parallel and in series with the UASBR, which was operated to develop its own granular sludge from wastewater treatment plant sludges (Owens 1994).

Analysis

On-site measurements and analyses include dissolved oxygen, nitrate, pH, electrical conductivity, and alkalinity. Samples are taken weekly for off-site laboratory analysis of selenium, total organic carbon (TOC), and solids. Samples are taken monthly for laboratory analysis of calcium, sodium, magnesium, potassium, chloride, sulfate, boron, chromium, molybdenum, uranium, and vanadium.

The selenium analyses and speciation procedures described in Standard Methods (3500-Se A, 17th edition) were adopted. This method involves continuous hydride generation, which depends on the reaction of selenite (+4 valence) with a strong reducing agent (sodium borohydride) to generate gaseous hydrogen selenide. The hydrogen selenide is swept into a heated quartz cell aligned in the optical path of a Varian SpectrAA-10 equipped with a vapor generation assembly (VGA-76) and a selenium lamp, and operating at 196 nm. The signal produced by the gaseous hydrogen selenide is proportional to the amount of selenite present in the sample. The detection limit for the analysis is 1 µg/L (1 ppb).

Total selenium is measured by first oxidizing all species to selenate (+6 valence) with potassium permanganate. The selenate is then reduced to selenite by reduction with hot hydrochloric acid. For the discussion of selenium results, total selenium refers to the concentration of all selenium species in the sample before filtration; soluble selenium is the total selenium in the sample after passing through a 0.22-μm laboratory membrane filter; selenite is measured without any sample pretreatment steps; and particulate selenium is the difference between the total and soluble selenium values.

Phase I Results – UASBR Trials with Imported Granular Sludge

A description of the initial startup and first 18 weeks of operations was described by Kipps (1994). In summary, the UASBR was filled with granular sludge and supernatant transported from a dormant UASBR at a Kansas City, Missouri, bakery. The settled granules occupied approximately 2 m^3 of the reactor space. After 1 month of acclimation and startup, the influent flowrate was set at 7.6 L/min and the recycle flow at 300 L/min. Methanol was added to the influent at a concentration of approximately 250 mg/L (94 mg/L total organic carbon). The reactor was operated for 32 weeks when it had to be shut down to repair a mechanical failure of the gas-solids separation cone and baffle. Most of the granular sludge had been lost by that time. Additional reactor modifications to improve operations were made at that time.

During this first 7-month period, the influent had a nitrate concentration of about 45 mg/L of NO_3^--N. Effluent nitrate concentrations were consistently low, usually between nondetectable and 3 mg/L NO_3^--N.

As shown on Figure 1, the granular sludge quickly acquired the ability to reduce selenate. At about 10 weeks after startup, when reactor temperature dropped below 15°C, the effluent-soluble selenium concentrations began to rise. However, even at 7°C – the lowest temperature recorded – 35% of the soluble selenium was being converted to a particulate form within the reactor. In week 22, additional mixing was provided to the reactor contents by diverting about 25% of the recycle flow through a mixing pipe inserted in the sludge bed. The effluent total selenium increased immediately as settled particulate selenium was resuspended. In addition, the soluble selenium removal increased to over 90%. Shortly after this removal was achieved, during the next few weeks a series of mechanical problems occurred, and the reactor was shut down for repairs in week 32.

Upon the shutdown of the reactor, it was discovered that a minimal amount of sludge remained in the reactor. It was speculated that the loss of sludge was due to the failure of the solid separation system. At the time, there was not an accurate method to determine the amount of sludge within the reactor.

After the reactor was repaired and modified with a sludge level monitoring system, it was reloaded with approximately 3.5 m^3 of the granular bakery sludge and restarted. Then 18 days later, a soluble selenium removal of 94% was achieved. However, even with modifications to the reactor, the granular sludge was being lost. For reasons still unclear, the sludge granules were able to navigate the system of internal baffles designed to retain them and float to the top of the reactor, where they were carried out in the effluent. Within 1 month of restarting, only 0.4 m^3 of sludge was left in the reactor and the test was stopped.

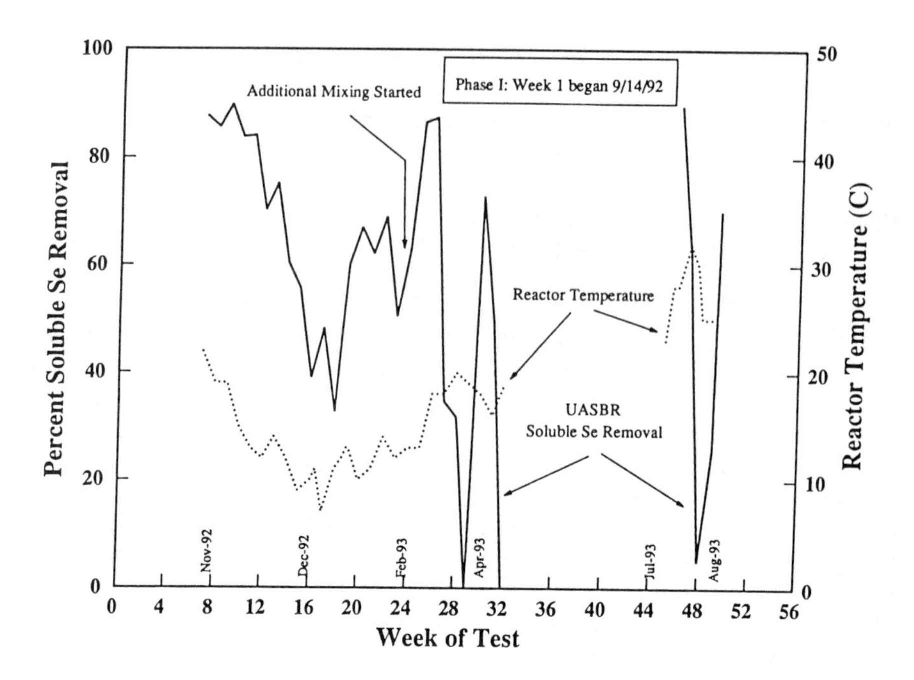

FIGURE 1. UASBR Se removal performance and reactor influent temperature during Phase I.

Because the granular sludge was not being retained and the possibility of getting large quantities of granular sludge for bigger systems would be limited, it was decided that a local source of biomass should be used to seed the reactor in Phase II.

Phase II Results — In Situ UASBR Sludge Formation and Testing

Phase II began with seeding the UASBR intermittently with anaerobic digester sludge, then activated sludge, both from local municipal sewage treatment facilities. A small quantity of sand was added to aid in the development of granular sludge with good settling characteristics. Figure 2 shows Phase II UASBR temperature and selenium-removal performance.

Prior to week 14, the UASBR upflow velocity ranged from 0.3 to 0.6 m/h and sludge volume varied from 0.8 to 2.3 m^3. Beginning in week 14, the reactor received 6.4 m^3 of activated sludge, and a recycle flow was initiated to increase the upflow velocity to 2.5 m/h. By week 16 most of the flocculent sludge exited with the effluent, and the bed volume increased by less than 0.5 m^3. In week 22, the upflow rate was lowered to 0.5 m/h by reducing the recycle flow and maintaining the inflow rate at 18.9 L/min. The sludge gradually increased to 3 m^3, or 25% of total reactor volume by week 40. From weeks 22 to 40, denitrification and selenium-removal performance improved with increasing influent temperature,

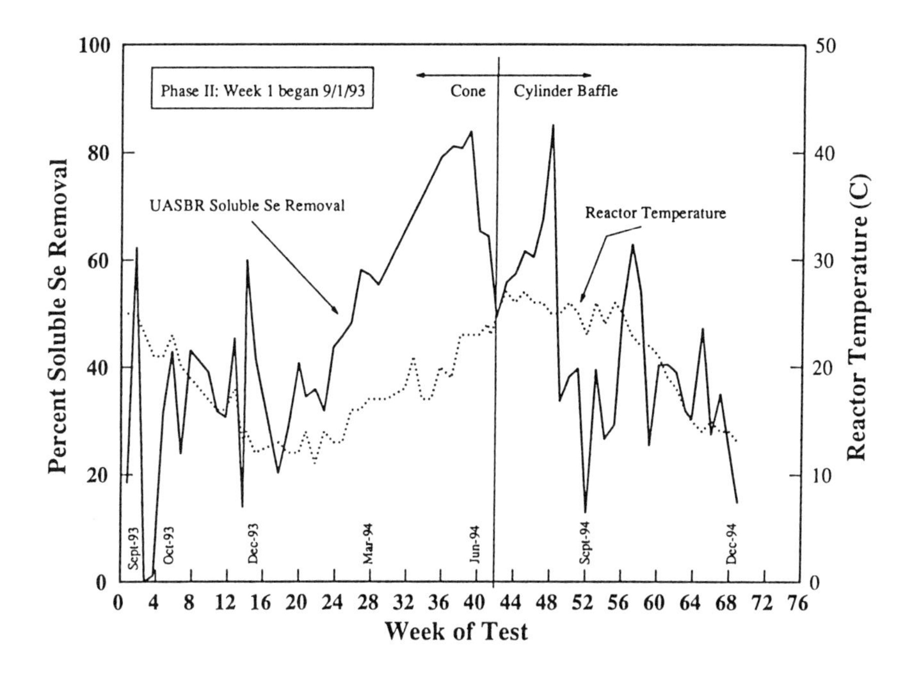

FIGURE 2. UASBR Se removal performance and reactor influent temperature during Phase II.

with complete denitrification being achieved in week 40. In week 42, several operational changes were made to the UASBR. The influent flowrate was doubled while maintaining the upflow velocity at 0.5 m/h and the reactor was operated to maintain a low nitrate residual. Also in week 42 the gas collection cone failed and was replaced with a cylinder baffle. These operation parameters were maintained for the rest of the year.

As in Phase I, methanol was added to achieve denitrification and selenium reduction. UASBR selenium-removal rates relative to total reactor volume ranged from 0.2 to 2 g Se/$m^3 \cdot$ d. Periodic analyses of UASBR sludge revealed a steady increase in inorganic solids (up to 80%) and selenium concentrations of up to 260 mg/kg.

UPCOMING RESEARCH ACTIVITIES

Continued operation of the UASBR will focus on optimizing operating conditions for selenium-removal performance, along with monitoring the production and composition of the biological reactor gas. Additional work will be performed on enhancing the operating parameters to encourage the growth and activity of indigenous selenium-reducing bacteria (e.g., Macy et al. 1993). Alternative sources of carbon, such as food-processing waste, will be identified and tested to replace methanol for denitrification and selenium removal.

REFERENCES

EPOC Ag. 1987. *Removal of Selenium from Subsurface Agricultural Drainage by an Anaerobic Bacterial Process.* Submitted to California Department of Water Resources, Fresno, CA.

Kipps, J. L. 1994. "Bioremediation of Selenium Oxides in Subsurface Agricultural Drainage Water." In J. Means and R. Hinchee (Eds.), *Emerging Technology for Bioremediation of Metals.* Lewis Publishers, Boca Raton, FL.

Macy, J. M., S. Lawson, and H. DeMoll-Decker. 1993. "Bioremediation of Selenium Oxyanions in San Joaquin Drainage Water Using *Thauera selenatis* in a Biological Reactor System." *Applied Microbiology and Biotechnology*, *40*:588-594.

Owens, L. P. 1992. *Adams Avenue Agricultural Drainage Research Center — Status Report, March 1993.* Submitted to California Department of Water Resources, Fresno, CA.

Owens, L. P. 1994. *Adams Avenue Agricultural Drainage Research Center — Progress Report on Reactor Operations for the Period September 1992 — July 1994.* Submitted to California Department of Water Resources, Fresno, CA.

Bacterial Bioremediation of Selenium Oxyanions Using a Dynamic Flow Bioreactor and Headspace Analysis

Steven L. McCarty, Thomas G. Chasteen, Verena Stalder, and Reinhard Bachofen

ABSTRACT

The volatile products of the biological reduction and methylation of selenium's most common oxyanions, selenate and selenite, were determined using capillary gas chromatography and fluorine-induced chemiluminescence detection. Dimethyl selenide and dimethyl diselenide were detected in the headspace above cultures of bacteria resistant to this metalloid using static and dynamic headspace sampling techniques. Fluorine-induced chemiluminescence detection was applied to determine the relative concentrations of the organosulfur and organoselenium species released over many days of culture growth at a controlled temperature and purge rate (for dynamic headspace sampling). A selenium-resistant bacterium, *Pseudomonas fluorescens* K27, and a phototrophic bacterium *Rhodobacter sphaeroides* 2.4.1 were exposed to SeO_4^{2-}, and the cultures' headspaces were examined over a period of several days for volatile selenium-containing products. The results show that the relative production of the volatile species over time depicts a pattern generally independent of the growth phase in the case of the phototrophic bacterium; the concentrations of metabolic dimethyl sulfide and dimethyl selenide determined in static headspace were highest after the microbe had been in stationary phase for 4 days. For *P. fluorescens*, bioremediation activity peaked soon after the end of the log phase (about 10 hours) and continuously decreased over the next 80 hours of this dynamic headspace experiment.

INTRODUCTION

The determination of the products of biodegradation can help us to understand the mechanisms of bioremediation and parameters necessary to optimize this process. Because of the chemical components normally present in the headspace above bacterial cultures, determining the organoselenium components produced by the reduction and methylation of selenium oxyanions requires a high-resolution separation technique and a specialized detector. The flame ionization detector (FID) responds very strongly to hydrocarbons usually present in very high concentrations in these systems (Kiene et al. 1986; Oremland & Zehr 1986) which complicates or prevents the analysis of the selenium-containing species at lower concentrations. Capillary gas chromatography partially solves this problem with its high-resolution abilities, but the high concentrations of water and/or CO_2 present can extinguish the flame or distort FID chromatography.

The selective detection abilities of fluorine-induced chemiluminescence are well suited to the determination of organosulfur, organoselenium, and organotellurium produced by biodegradation of oxyanions of these toxic metalloids (Chasteen et al. 1990; McCarty et al. 1993; Stalder et al. 1995). Thus, two headspace sampling methods were used to determine the change with time in relative concentrations of the volatile products above (1) an anaerobic culture of a selenium-resistant bacteria isolated at Kesterson Reservoir (Burton et al. 1987), *Pseudomonas fluorescens* K27, a facultative anaerobe; and (2) a phototrophic nonsulfur bacterium, *Rhodobacter sphaeroides* 2.4.1, which has also shown resistance to selenate and selenite (Moore & Kaplan 1992; McCarty et al. 1993). Kesterson Reservoir is a site containing relatively high concentrations of dissolved selenium, mostly as selenate and selenite oxyanions (Ingersoll et al. 1990).

EXPERIMENTAL PROCEDURES AND MATERIALS

Static Headspace Experiments

All of the static headspace experiments used purple nonsulfur bacterium *Rhodobacter sphaeroides* 2.4.1. (DSM # 158). The cells were grown on Sistrom minimal medium (SMM) (Sistrom 1960) with 20 mM succinate as the sole carbon source. After inoculation of the liquid medium, the cultures were left in the dark overnight before being incubated in incandescent light (10 Wm^2; 100 W tungsten light bulb) at 28°C. With these microbes, anaerobic growth is quickly achieved after O_2 is used up in aerobic (nonphotosynthetic) growth in the dark.

A 45-mL culture of the phototroph was grown in a 100-mL Schott® flask sealed with a specially designed enclosure cap (Stalder et al. 1995). This device allows for the repeated sampling of the gas and solution in the flask without exposing the culture to O_2. Before adding selenium (1 mM final concentration), this culture was incubated in the dark overnight and then in the light for 2 to 3 days. Sodium selenate (Na_2SeO_4) in 15 mL of 4 times concentrated SMM was added to bacterial cultures from a sterile stock solution. The culture was opened for the selenium amendment, sealed, and again left in the dark overnight to become anaerobic before it could be exposed to light. Na_2SeO_4 was obtained from Strem Chemicals (Newburyport, Massachusetts). Gas sampling was accomplished using a gastight syringe. The culture solution was removed by a disposable syringe.

The optical density of phototrophic bacterial cultures at 660 nm as a measure for biomass and the absorption at 875 nm for the relative amount of light-harvesting complex were determined on a Bausch and Lomb (Rochester, New York) Spectronic 710 spectrophotometer.

Dynamic Headspace Experiments

The dynamic headspace reactor, cryogenic trapping system, and gas chromatographic (GC) system are schematically depicted in Figure 1. Although the nominal volume of the glass bioreactor was 250 mL, the headspace depended on the volume of liquid medium contained. This experiment used 150 mL of the complex medium, tryptic soy broth (10 g/L; Difco, Detroit, Michigan), with 1 g/L sodium nitrate added as electron acceptor (hereafter called TSN medium). To the sterile bioreactor containing sodium selenate in sterile TSN medium was added 20 mL of TSN medium containing K27 bacteria that had been grown into the stationary phase overnight; the final Na_2SeO_4 concentration was 10 mM. The reactor was closed and immediately purged with N_2 @ 3.0 mL/min and then maintained at 29 ± 2°C in a constant temperature water bath for the rest of the experiment.

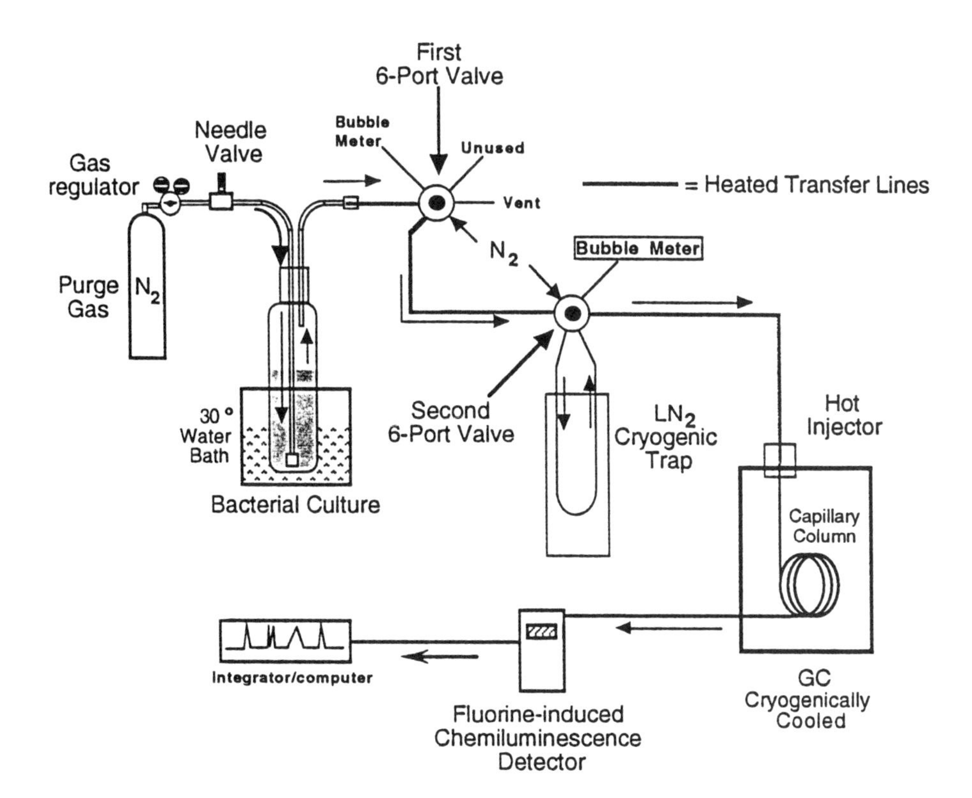

FIGURE 1. Schematic diagram of dynamic headspace sampling system.

The cryogenic trap consisted of 45 cm of 0.32-cm Silcosteel bent into a U to fit inside an insulated liquid nitrogen Dewar (37 cm long × 19 cm wide). During cryogenic trapping the Dewar was purged with liquid nitrogen (LN_2) cooled air in a separate vessel (not shown) to establish a temperature of approximately −60°C in the trap. At this point bioreactor headspace flowed through the first 6-port valve and into the cryogenic trap. Gas exiting the trap during this step flowed through the second 6-port valve and through a bubble meter to waste so that the gas-trapping flow could be monitored. When trapping was completed, the trap was heated to 180°C to evaporate the condensed components. For sample injection, the 6-port valves were rotated such that clean N_2 flowed through the trap and swept the vaporized analytes off of the trap and into the GC where they were trapped again at −20°C at the head of the column. Between runs all transfer lines were kept hot and purged with clean N_2 to prevent sample carryover. Periodic blank runs, with 10- to 20-minute trapping times of clean purge gas, were made to prevent sample carryover between runs.

Depending on the phase of growth and concentration of headspace components different gas sample trapping times were used. Since the purge gas flow through the bioreactor was constant at 3.0 mL/min of N_2, these different trapping times translate to different volume of headspace gas captured and chromatographed. All gas-phase, headspace concentrations are reported as parts per billion by volume (ppbv).

GC Analysis With Fluorine-Induced Chemiluminescence Detection

All chromatographic analyses were performed using capillary gas chromatography linked to a fluorine-induced chemiluminescence detector as described elsewhere (McCarty et al. 1993; Zhang & Chasteen 1994). In brief, a relatively nonpolar capillary column was used for chromatographic separation of all organosulfur and organoselenium components reported. Component identities and calibration were based on the retention times and integrated peak areas of chemical standards injected at known concentrations in reagent-grade acetonitrile. Helium was used as the carrier gas. For the syringe-injected samples (static headspace experiments), in-oven cryogenic trapping was accomplished by subambient cooling of the GC oven using LN_2 and the following temperature program: −20°C for 1 min initial time, 8°C/min to 20°C, then an immediate 15°C/min ramp to 200°C for 3 min. Samples were injected primarily in the splitless mode; however, split injections (split ratio 1/65) were performed occasionally for higher-concentration samples. Peak areas from split injections were normalized to areas of splitless injections by multiplying by 1/split ratio. Static headspace was sampled with a gastight syringe and immediate injection into the 275°C GC injector.

This detector's response is linear over 3 orders of magnitude from low picograms to approximately 25 nanograms on-column for all of the organosulfur and organoselenium compounds determined. The limits of detection for the compounds reported here are approximately 25 picograms at a signal-to-noise ratio equal to 3.

RESULTS

The results of a single time course experiment for *Rhodobacter sphaeroides* are shown in Figure 2. The experiment lasted approximately 180 hours. The top graph shows the culture's optical density (OD) at 660 nm as a measure of biomass and the ratio of optical density at 875 and 660 nm as a measure of the amount of photosynthetic units per biomass. The middle and bottom graphs of Figure 2 plot the *relative* variation in organoselenium and organosulfur components in the culture's headspace over time: dimethyl selenide (DMSe), dimethyl diselenide (DMDSe), dimethyl sulfide (DMS), and dimethyl disulfide (DMDS).

Figure 3 presents the time course results from the dynamic headspace experiment of a single anaerobic culture of *Pseudomonas fluorescens* determined for over 100 hours. The culture was purged continuously with N_2 at 3 mL/min for the entire experiment. The plot shows the relative headspace concentration of DMDS and DMDSe in ppbv over time. Data for the CH_3SH (methanethiol), DMS, and DMSe were taken from these chromatograms; however, they were much more variable, lower in concentration, and often slipped below the detection limit ($\approx$ 1 ppbv for 4 min trapping time). Because our experience with these bacteria has shown a relatively high concentration of CH_3SH and DMS in static headspace samples (Chasteen et al. 1990), the lower concentrations determined here suggest that either these lower boiling components were somehow lost during the trapping step or that the continuous N_2 purging of the headspace kept the concentration much lower than what was allowed to build up during static headspace experiments.

DISCUSSION

The data in Figure 2 for the phototrophs show that the highest concentrations of DMS and DMSe were found in the stationary phase of growth where some growth requirement has become limited to *R. sphaeroides*. This reduction and methylation therefore are powered only by light and not by reducing power produced in metabolizing and anabolizing cells.

The dynamic headspace experimental data displayed in Figure 3 represent the first experiment that we know of that tracks the production of a volatile bioremediation product from a single culture (*Pseudomonas fluorescens* K27) over a time as long as 100 hours. With the culture and its headspace continuously purged with nitrogen, the concentrations of volatile organosulfur or organoselenium species were not allowed to build up in the culture medium or headspace as in our previous static headspace culture experiments (Chasteen et al. 1990; McCarty et al. 1993; Zhang & Chasteen, 1994); therefore the metabolic processes of these bacteria probably were not affected by the volatile selenium-containing species produced.

Even when stressed by up to 50 mM selenate, newly exposed cultures of this bacterium achieve stationary phase in approximately 10 hours (data not shown).

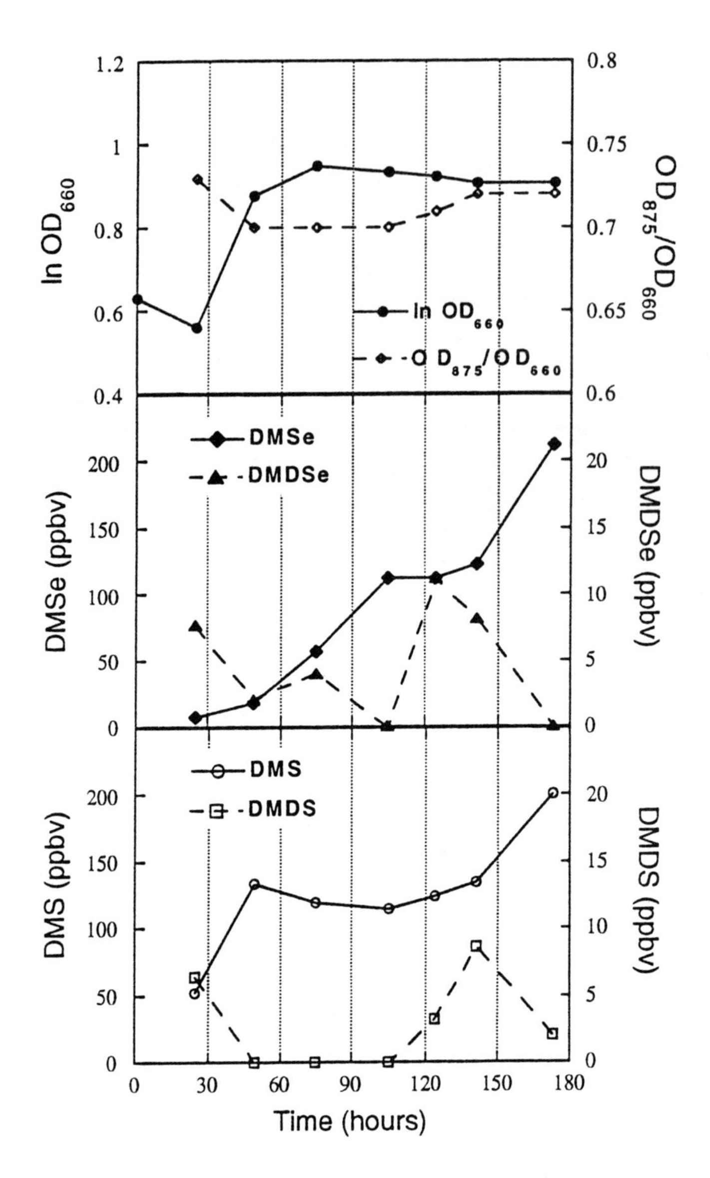

FIGURE 2. Static headspace time course experiment with *Rhodobacter sphaeroides* 2.4.1 amended with 1 mM selenate. OD is optical density at wavelength specified.

The production of the DMDS and DMDSe shows a maximum after the achievement of stationary phase and a gradual decrease in the relative headspace concentrations of these species over time. Unlike the phototrophic bacteria examined in the static headspace experiments described above, with K27 the reduction and methylation of selenium appears to be at maximum at or near the end of the log phase of growth.

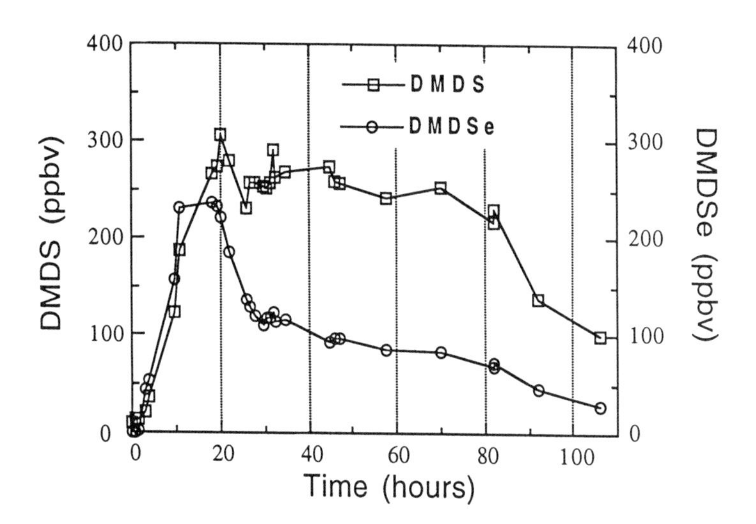

FIGURE 3. Dynamic headspace time course experiment with *Pseudomonas fluorescens* K27 amended with 10 mM selenate.

ACKNOWLEDGMENT

This research was supported by an award from Research Corporation.

REFERENCES

Burton, G. A., T. H Giddings, P. De Brine, and R. Fall. 1987. "High incidence of selenite-resistant bacteria from a site polluted with selenium." *Appl. Environ. Microbiol. 53*(1): 185-188.

Chasteen, T. G., G. M. Silver, J. W. Birks, and R. Fall. 1990. "Fluorine-induced chemiluminescence detection of biologically methylated tellurium, selenium, and sulfur compounds." *Chromatographia 30*(3/4): 181-185.

Ingersoll, C. G., F. J. Dwyer, and T. W. May. 1990. "Toxicity of Inorganic and organic selenium to *Daphnia magna* (Cladocera) and *Chironomus riparius* (Diptera)." *Environ. Toxicol. Chem. 9*: 1171-1181.

Kiene, R. P., R. S. Oremland, A. Catena, L. G. Miller, and D. G. Capone. 1986. "Metabolism of reduced methylated sulfur compounds in anaerobic sediments and by a pure culture of an estuarine methanogen." *Appl. Environ. Microbiol. 52*(5): 1037-1045.

McCarty, S. L., T. G. Chasteen, M. Marshall, R. Fall, and R. Bachofen. 1993. "Phototrophic bacteria produce volatile, methylated sulfur and selenium compounds." *FEMS Lett. 112*: 93-98.

Moore, M., and S. Kaplan. 1992. "Identification of intrinsic high-level resistance to rare-earth oxides and oxyanions in members of the class proteobacteria: Characterization of tellurite, selenite, and rhodium sesquioxide reduction in *Rhodobacter sphaeroides*." *J. Bact. 174*(5): 1505-1514.

Oremland, R. S., and J. P. Zehr. 1986. "Formation of methane and carbon dioxide from dimethylselenide in anoxic sediments and by a methanogenic bacterium." *Appl. Environ. Microbiol. 52*(5): 1031-1036.

Sistrom, W. R. 1960. "A requirement for sodium in the growth of *Rhodopseudomonas sphaeroides*." *J. Gen. Microbiol.* 22: 778-785.

Stalder, V., N. Bernard, K. W. Hanselmann, R. Bachofen, and T. G. Chasteen. 1995. "A method of repeated sampling of static headspace above anaerobic bacterial cultures with fluorine-induced chemiluminescence detection." *Anal. Chim. Acta* *303*(1): 91-97.

Zhang, L., and T. G. Chasteen. 1994. "Amending cultures of selenium-resistant bacteria with dimethyl selenone." *Appl. Organomet. Chem.* *8*(6): 501-508.

Volatilization of Arsenic Compounds by Microorganisms

Reinhard Bachofen, Linda Birch, Urs Buchs, Peter Ferloni, Isabelle Flynn, Gaudenz Jud, Harald Tahedl, and Thomas G. Chasteen

ABSTRACT

Methanobacterium thermoautotrophicum is resistant to arsenate; concentrations of up to 350 μM had no effect on growth or methanogenesis. Continuous cultures fed with arsenate transformed arsenic (As), depending on the phosphate concentration, to volatile arsenic species with a yield of 25%. Similarly, various phototrophic purple bacteria are resistant to arsenate and other metalloidal oxyanions at concentrations up to 30 mM; they are able to reduce arsenate. When soil contaminated with arsenic was leached in percolation columns under various environmental conditions, indigenous microorganisms solubilized and slightly volatilized the arsenic present.

INTRODUCTION

Arsenic compounds are universally distributed in ultralow concentrations in food, soil, and water; the element may be locally accumulated due to human activities such as mining of metals or the production of pesticides (Schraufnagel 1982). Arsenic is present in nature in a variety of chemical species (Andreae 1982; Cullen & Reimer 1989), some of them extremely toxic to living organisms, such as arsenate, arsenite, or arsenic trioxide. The latter has been used extensively throughout history to poison humans. Furthermore arsenic compounds still create interest and have been used as the active reagents in chemical warfare, e.g., Lewisite. They have also been used in the tanning industry and are part of wood conservation chemicals; some are important as herbicides. Finally, large amounts of As are brought into the environment by burning coal.

Arsenic-contaminated soil often is treated by washing or heating. Volatilization by fungi or bacteria, and accumulation in plants and algae have been suggested as methods of bioremediation. We present experiments on bioleaching of contaminated soil by a bacterial mixed culture and volatilization of soluble As species by a methanogenic culture.

MATERIALS AND METHODS

Soil material contaminated with As was obtained from the site of an abandoned tannery. In soil profiles, As concentrations increased from about 50 mg/kg dry soil in the upper 0 to 100 cm up to 24 g/kg dry soil in the layer between 120 and 180 cm in depth. Sulfur showed a similar profile suggesting the presence of As as As_2S_2. The homogenized soil (air-dried) used for the leaching experiments contained 8.26 g As/kg. In a second experiment, fresh garden soil was spiked with As_2S_3 (1.36 g As_2S_3/100 g air-dried soil, equal to 8.26 g As/kg as above).

Total arsenic was determined by inductively coupled plasma-atomic emission spectroscopy (ICP-AES, detection limit: 1 mg As/L, Spectro Analytical Instruments) directly or after wet digestion (Microwave combustion: MLS 1200, with HNO_3 and H_2O_2) or by hydride generation atomic absorption spectroscopy (HGAAS, detection limit: 10 µg As/L, Perkin-Elmer, AAAS-4000). While ICP-AES gives the total concentration of As independent of the As-containing species present, with HGAAS the signal size is dependent on the species composition; thus for each species the instrument has to be calibrated.

Soluble arsenic oxyanions were separated either by ion chromatography (Metrohm 690 Ion chromatograph, column Hamilton PRP-x100, 125 × 4mm) or by capillary electrophoresis (CE, Grom, system 100). A typical separation of As compounds by CE is shown in Figure 1. For speciation, the volatile As compounds were trapped in heptane (–80°C) and analyzed by capillary GC-MS (HP-G1800A, column HP-5, 30 m 0.25 mm i.d., 50°C, carrier gas He, 0.5 mL/min).

Methanobacterium thermoautotrophicum was cultivated in a 2-L bioreactor (MBR, Wetzikon, Switzerland) stirred at 1200 rpm at 60°C and pH 6.9, under controlled chemostat conditions, using H_2 and CO_2 as substrates for growth (Jud et al. 1995). The flowrates were 400 mL/min, 110 mL/min and 40 mL/min for H_2, CO_2, and Ar (for calibration), respectively. The concentration of methane in the exhaust gas was determined by infrared absorption. Volatile As compounds were trapped in 4 sequential vessels cooled in ice containing 40% HNO_3 to oxidize and concentrate the species evolved for total As and in cold heptane (–80°C) for the analysis of the volatile species.

Pure cultures of phototrophic purple bacteria (*Rhodospirillum rubrum* S1, *R. rubrum* G9, *Rhodobacter sphaeroides* 158, *R. sphaeroides* 2340, *Rhodocyclus tenius*, *Rhodopseudomonas blastica*, and *Rhodobacter capsulatus* were grown in the medium described by Sistrom (1960) at 30°C either in closed, completely filled 100 mL bottles or in Petri dishes in anaerobic jars in the light.

Leaching experiments were performed with soil columns (amount of dry soil 100 g, liquid volume 700 mL) with a percolation system modified from Audus (1946) and Lees and Quastel (1946); the liquid containing a dilute nutrient solution (glucose 1 g/L, $NH_4.HCO_3$ 0.34 g/L) was circulated at a rate of 60 mL/h. Liquid samples were taken for analysis manually with a syringe. The whole system was purged continuously with nitrogen, and volatile species of As were collected in 3 serially placed gas traps, each containing 70 mL of 40% HNO_3 w/w (Suprapur, Merck).

RESULTS

Resistance of Pure Bacterial Cultures towards AsO_4^{-3} and AsO_3^{-3}

M. thermoautotrophicum was able to grow without change in physiological activity as measured by the rate of methane formed up to concentrations of 350 M arsenate.

Growth of the phototrophic purple bacteria such as *Rhodospirillum rubrum* was not influenced by arsenate up to concentrations of 30 mM at high-light conditions (14 mW) and up to 150 mM in low-light conditions (7 mW). In contrast, arsenite became growth inhibiting and induced lag phases up to 200 h at concentrations of 3 mM in low light conditions. Many phototrophic organisms have been shown previously to be resistant towards a variety of oxyanions; thus it

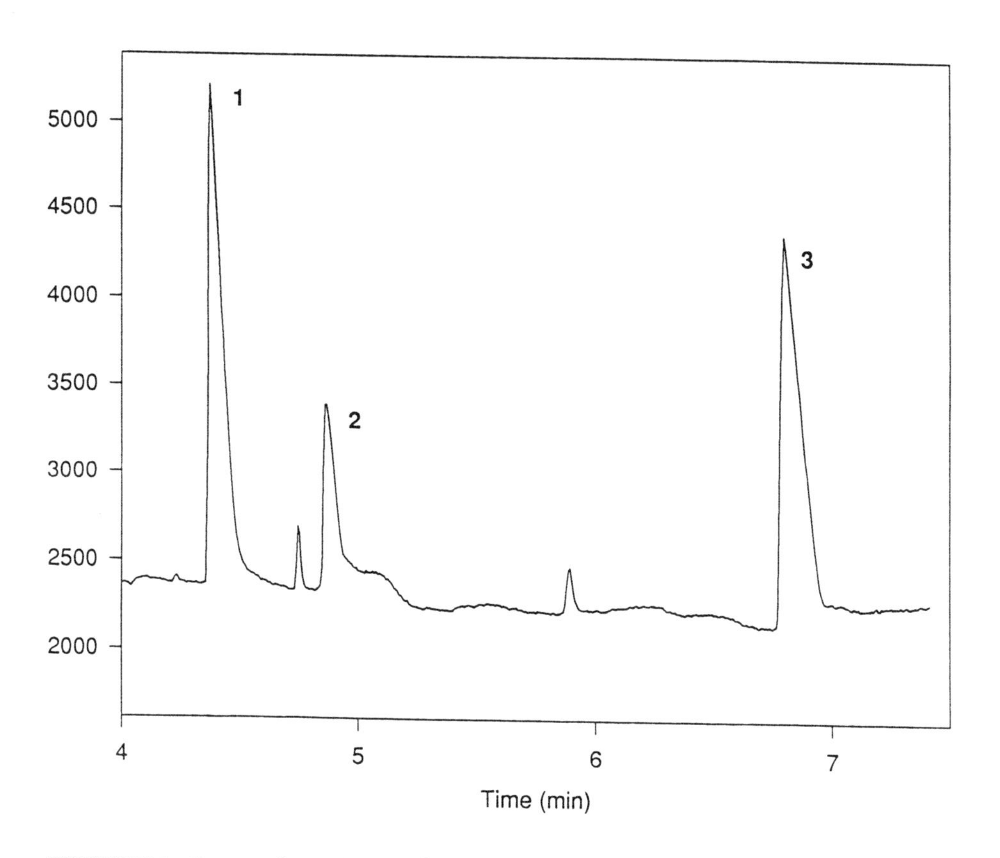

FIGURE 1. Separation of AsO_4^{-3} (1), AsO_3^{-3} (2), and $(CH_3)_2AsO_2^-$ (3) by capillary electrophoresis. Buffer: Na_2CrO_4 (5 mM) and OFM-Anion BT (0.5 mM) Waters, USA, pH 8.0, capillary 50 cm, 50 μm i.d., voltage 20 kV, output signal in μV, range 0.05. Detection limit for the 3 compounds: 0.1 ppm.

seems that tolerance towards these toxic elements is widely distributed among phototrophs (Moore & Kaplan 1992).

Leaching of Soil Columns Containing As

Arsenic-contaminated soil was percolated with nutrient medium under aerobic and anaerobic conditions. Growth of microorganisms, as measured by the increase in turbidity, was high in the presence of O_2, but needed an incubation time of a few days to get started in the absence of O_2. Controls sterilized with formaldehyde showed no increase in turbidity. However 2 to 3 times more As was volatilized under anaerobic conditions as detected in the exhaust gases. Garden soil spiked with As_2S_3 therefore was leached under anaerobic conditions.

Within 50 days, the As concentration in the leaching medium increased up to 580 ppm. In the sterile control (soil poisoned with 0.05% formaldehyde), less than 10 ppm soluble As was found at the end of the experiment as a result of a purely physicochemical leaching process (Figure 2). The percentage of As extracted was close to 50% after this period. In the acid traps, small amounts of volatile As compounds (13 µg) were found after as little as 3 days; volatile As emissions remained constant during the following 30 days and then increased considerably to 45 µg. After 50 days, however, only about 0.01% of the As added to the soil was found in the gas traps.

Formation of Volatile Compounds by *M. thermoautotrophicum*

Methanogens are known to methylate arsenate (McBride & Wolfe 1971). In our experiments, volatile As compounds were trapped in the exhaust gases from the bacterial culture under stable chemostat conditions when fed with 5 to 350 µM arsenate. Compared to the control lacking arsenate, no significant change in biomass concentration, neither in the rate of methane formation nor in the size of the redox potential in the culture vessel, was observed. Volatilization of As was stimulated by lowered phosphate concentrations, which had no effect upon the growth of the cells in a concentration range from 0.75 mM to 10.0 mM. Under our conditions, growth was still limited by the supply of hydrogen to the cells.

By increasing the ratio of added arsenate to phosphate present, the conversion efficiency into volatile species could be increased to 25% under the continuous culture conditions described. In our experiments *M. thermoautotrophicum* reduced the arsenate in the medium mainly to arsine (AsH_3) with small amounts of dimethylarsine ($HAs(CH_3)_2$).

DISCUSSION

Biotransformations of As have been found in a variety of natural ecosystems (Andreae 1982; Cullen & Reimer 1989); As may thus be present as As^V, As^{III},

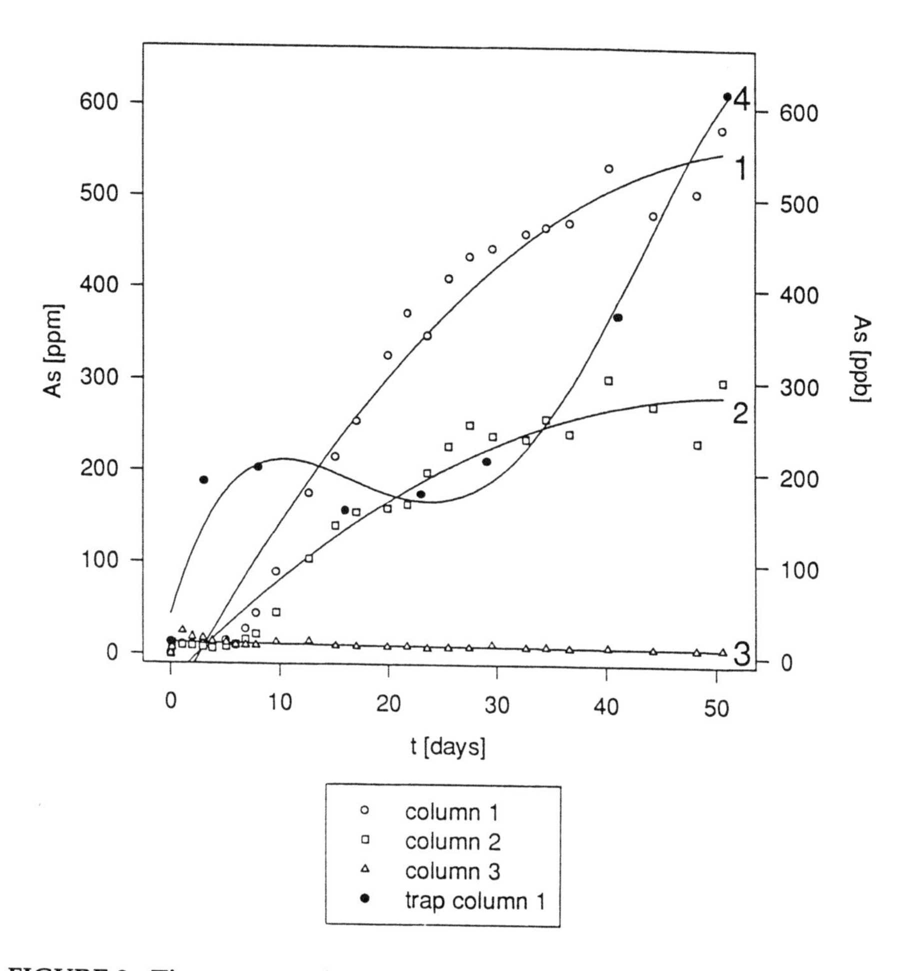

FIGURE 2. Time course of As leaching from garden soil spiked with As_2S_3. (1) and (2) water soluble As in two different leaching systems and (3) in the sterile control, (4) As found in traps (70 mL) from leaching system (1) above. Units are given in ppm for (1)-(3) and in ppb for (4).

and in methylated forms, mainly depending on the presence of oxygen. The oldest report of toxic volatile compounds from As-containing wall-paper stems from Gmelin in the first half of the 19th century, an observation further studied by Gosio (1897). The gas formed under aerobic conditions was later identified as trimethylarsine (Challenger et al. 1933) while in our experiments with *M. thermoautotrophicum* the main volatile As species present was arsine. A variety of microorganisms is able to reduce AsO_4^{3-} to AsO_3^{-3}. AsO_4^{-3} may even be an electron acceptor in anaerobic respiration (Ahmann et al. 1994).

Microorganisms that solubilize As by oxidation from insoluble minerals have been described (Tuovinen et al. 1994). In our leaching experiments, an undefined bacterial population converted the insoluble As_2S_3 to nearly 50% water-soluble form within 2 months. The experiments show that bacterial leaching of material contaminated with As could be successfully used for bioremediation. Using an anaerobic reactor inoculated with methanogenic bacteria, a technical system might be developed in which As is extracted from the solid phase in a volatile form that could be purged, trapped, and collected.

REFERENCES

Ahmann, D., A. L. Roberts, L. R. Krumholz, and F.M.M. Morel. 1994. "Microbes grow by reducing arsenic." *Nature 371*: 750.

Andreae, M. O. 1982. "Biotransformation of arsenic in the marine environment." In: W. H. Lederer and R. J. Fensterheim (Eds.), *Arsenic: Industrial, Biomedical, Environmental Perspectives.* Van Nostrand Reinhold Company, New York, NY. pp. 378-392.

Audus, L. J. 1946. "A new soil perfusion apparatus." *Nature 158*: 419.

Challenger, F., C. Higginbottom, and L. Ellis. 1933. "The formation of organo-metalloidal compounds by microorganisms. I. Trimethylarsine and dimethylethylarsine." *J. Chem. Soc. 1933*: 95-101.

Cullen, W. R., and K. J. Reimer. 1989. "Arsenic speciation in the environment." *Chem. Revs. 89*: 713-764.

Gosio, B. 1897. "Zur Frage, wodurch die Giftigkeit arsenhaltiger Tapeten bedingt wird." *Ber. Deutsch. Chem. Ges. 30*: 1024-1026.

Jud, G., K. Schneider, and R. Bachofen. 1995. "The importance of the hydrogen mass transfer on the growth kinetics in batch and chemostat cultures of *Methanobacterium thermoautotrophicum.*" Submitted.

Lees, H., and J. H. Quastel. 1946. "Biochemistry of nitrification in soil." *Biochem. J. 40*: 803-823.

McBride, B. C., and R. S. Wolfe. 1971. "Biosynthesis of dimethylarsine by Methanobacterium." *Biochemistry 10*: 312-317.

Moore, M. D., and S. Kaplan. 1992. "Identification of intrinsic high-level resistance to rare-earth oxidesand oxyanions in members of the class *Proteobacteria*. Characterization of tellurite, selenite, and rhodium sesquioxide reduction in *Rhodobacter sphaeroides.*" *J. Gen. Microbiol. 22*: 778-785.

Schraufnagel, R. A. 1982. "Arsenic energy sources, a future supply or an environmental problem." In: W. H. Lederer and R. J. Fensterheim (Eds.), *Arsenic: Industrial, Biomedical, Environmental Perspectives.* Van Nostrand Reinhold Company, New York. pp. 17-41.

Sistrom, W. R. 1960. "A requirement for sodium in the growth medium of *Rhodopseudomonas sphaeroides.*" *J. Gen. Microbiol. 22*: 77-85.

Tuovinen, O. H., T. M. Bhatti, J. M. Bigham, K. B. Hallberg, O. Garcia, and E. B. Lindstrom. 1994. "Oxidative dissolution of arsenopyrite by mesophilic and moderately thermophilic acidophiles." *Appl. Environ. Microbiol. 60*: 3268-3274.

In Situ Biodenitrification of Nitrate Surface Water

Glen C. Schmidt and M. Bruce Ballew

ABSTRACT

The U.S. Department of Energy's Weldon Spring Site Remedial Action Project has successfully operated a full-scale in situ biodenitrification system to treat water with elevated nitrate levels in abandoned raffinate pits. Bench- and pilot-scale studies were conducted to evaluate the feasibility of the process and to support its full-scale design and application. Bench testing evaluated variables that would influence development of an active denitrifying biological culture. The variables were carbon source, phosphate source, presence and absence of raffinate sludge, addition of a commercially available denitrifying microbial culture, and the use of a microbial growth medium (BNB-MICRO+, Westbridge). Nitrate levels were reduced from 750 mg/L NO_3-N to below 10 mg/L NO_3-N within 17 days. Pilot testing simulated the full-scale process to determine if nitrate levels could be reduced to less than 10 mg/L NO_3-N when high levels are present below the sludge surface. Four separate test systems were examined along with two control systems. Nitrates were reduced from 1,200 mg/L NO_3-N to below 2 mg/L NO_3-N within 21 days. Full-scale operation has been initiated to denitrify 900,000-gal batches alternating between two 1-acre (0.41-hectare) ponds. The process used commercially available calcium acetate solution and monosodium/disodium phosphate solution as a nutrient source for indigenous microorganisms to convert nitrates to molecular nitrogen and water.

INTRODUCTION

The Weldon Spring Site (WSS) is a U.S. Department of Energy (U.S. DOE) remediation project located 30 miles west of St. Louis, Missouri. The WSS was formerly used to process uranium and thorium ore concentrates for weapons use. The site was abandoned in the late 1960s. In 1986 it was determined that the site required remediation under the Comprehensive Environmental Response, Compensation, and Liability Act (CERCLA). The site was also placed on the

National Priorities List (NPL). Several contaminants require treatment at the WSS, however, the focus of this paper is on the nitrate contained in surface water impoundments at the site.

Nitric acid was used in processing uranium and subsequently was discarded in the wastestream. The wastestream, raffinate, was discharged to one of four surface impoundments (raffinate pits), where the pH was adjusted with lime to precipitate the metals. The raffinate sludge generated from this process remains in the pits today. The chemical characteristics of each pit are different. The primary difference is the concentration of nitrate. Two of the impoundments, raffinate pits 1 and 3, have nitrate concentrations that require treatment prior to discharge per a National Pollutant Discharge Elimination System (NPDES) permit.

Unusually high precipitation in 1993 caused the freeboard levels in raffinate pit 3 to increase to a level threatening overtopping and subsequent uncontrolled release. This high-nitrate water could not be treated in the existing wastewater treatment facilities. Therefore, alternatives to treat raffinate pit 3 water had to be identified to reduce freeboard levels. In situ biodenitrification was identified as a potential alternative to reduce the nitrate concentration and thereby make the water candidate for treatment in the existing treatment facilities.

EXPERIMENTAL PROCEDURES

Bench-scale tests (SITEX Environmental Inc.) were conducted in six 7-L glass beakers. Beaker studies were used to evaluate carbon source, phosphate source, presence or absence of sludge from pit 3, addition of BI-CHEM DC 1008 SF (Sybron Chemical Co.), a commercially available denitrifying microbial culture, and the use of a biological stimulant (BNB-MICRO+, Westbridge). Two carbon sources evaluated were calcium acetate and tributylphosphate (TBP) at a carbon:nitrogen ratio of 1.4:1 (EPA). TBP ($C_{12}H_{27}O_4P$) is available on site and has the potential to provide the carbon and phosphate source. The phosphorus source used was a granular phosphate fertilizer (P_2O_5), dosed at 50 mg/L. Denitrifying bacteria were applied at 3 g/L. The microbial stimulant was dosed at 20 mg/L. Mixing was provided by recirculating the contents with a peristaltic pump. The following parameters were measured to evaluate each system: nitrate, dissolved oxygen, pH, temperature, total organic carbon, and solids production (by observation).

Pilot-scale testing (U.S. DOE 1994a) was conducted with larger 136-L test systems. Each system contained 30 L of raffinate pit 3 sludge and 110 L of water. Calcium acetate was added at a carbon:nitrogen ratio of 1.4:1. A mono/disodium phosphate solution was added to a concentration of 50 mg/L. One system received BI-CHEM DC 1008 SF culture. The following parameters were monitored throughout the test: nitrate, dissolved oxygen, pH, temperature, total organic carbon, sludge level, and water level. Phosphate was measured 9 days into the test.

RESULTS

The bench- and pilot-scale studies were successful in identifying the following relevant process parameters and engineering data for the subsequent pilot studies. Calcium acetate test systems reduced nitrate levels from 750 mg/L NO_3-N to 7 mg/L NO_3-N in 10 days. TBP test systems were unsuccessful at reducing nitrate levels. Systems that contained raffinate pit 3 sludge showed the greatest nitrate reduction in the shortest amount of time. The beaker test concluded that calcium acetate functions well as the carbon source for indigenous organisms present in raffinate sludge. A carbon:nitrogen weight ratio of 1.4:1 is sufficient for nitrates to be reduced below the design goal of less than 10 mg/L NO_3-N. There was some difficulty dissolving the granular pellets of the phosphate fertilizer. A liquid source of phosphate is recommended for additional applications.

Nutrient Requirements

The treatability study demonstrated that the best carbon source for our situation is calcium acetate. A carbon:nitrogen ratio of 1.4:1 was sufficient to maintain the denitrification reaction with minimal residual organic carbon remaining. Monosodium/disodium phosphate dosed to 50 mg/L was determined to be the best phosphate source for scaleup operations.

Dissolved Oxygen, pH, Temperature

The denitrification rate was directly related to the dissolved oxygen (DO) concentration. Measurable denitrification did not occur when the DO concentration was above 5 mg/L. The rate of denitrification increased significantly after the DO decreased to below 2 mg/L. The relationship of DO to denitrification was strong enough to use DO concentration as a primary indicator for optimum process conditions. The pH increased from neutral to approximately 1 to 2 standard units as the reaction proceeded. Although detailed studies were not performed to evaluate temperature effects, the denitrification rate significantly decreased when the ambient temperature was below 50°F (10°C). This observation is generally supported by available literature (U.S. EPA, 1993).

Raffinate Sludge and Microbial Seeding

The presence of the raffinate sludge, which consists primarily of various heavy metals and radionuclides, did not appear to have any inhibitory effects on the process. The nitrate found beneath the raffinate sludge surface did not leach into the surface water in the bench or pilot study. The addition of a commercially available denitrifying biological culture (BI-CHEM DC 1008 SF) did not measurably improve the process. This lack of improvement suggests that the denitrifying bacteria were already present in the raffinate pit surface water.

Figure 1 presents the data collected for one of the 136-L test systems. The data are representative of the data collected for each test system.

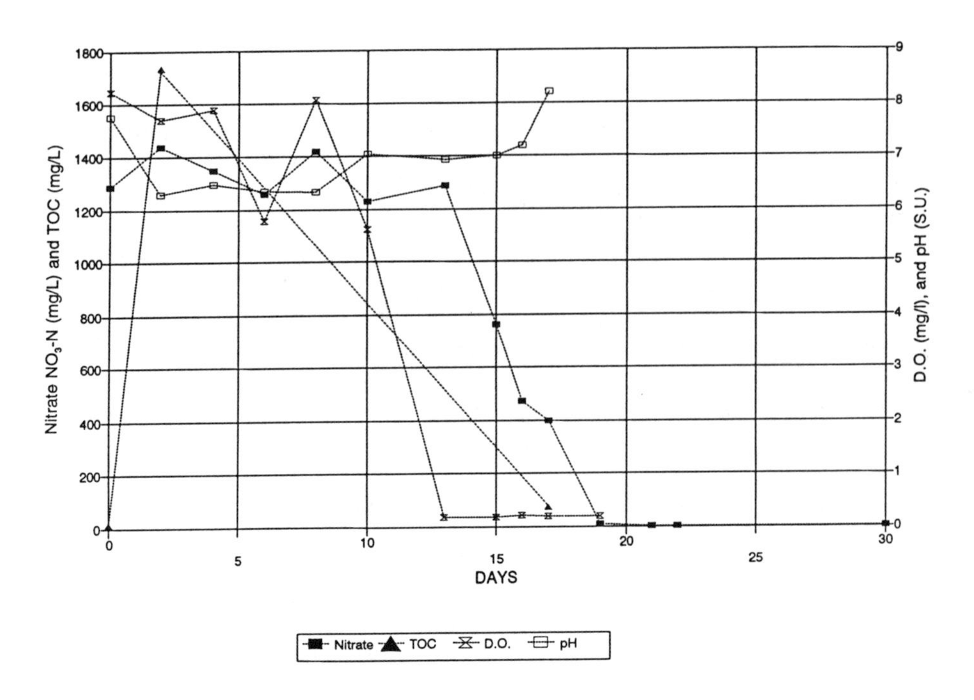

FIGURE 1. Pilot test system, nitrate reduction trend.

FULL-SCALE OPERATION

Raffinate pit 1 was selected for full-scale operation due to its relatively small volume (900,000 gals), ease of implementation and process control and minimal capital expenditure required. Information generated from the pilot study was used to develop the full-scale engineering criteria. Two 5,000-gal (18,927-L) tankers of calcium acetate were ordered from a bulk chemical supplier. These tankers were located adjacent to raffinate pit 1. Drums (55-gal [208-L]) of a mono/disodium phosphate (BiChem Accelerator IV, Sybron Chemical Co.) were also located adjacent to raffinate pit 1. The piping, fitting, and valves consisted primarily of polyvinyl chloride. The calcium acetate and phosphate injected into a recirculation line for dilution into the pit. The recirculation line was pumped via a gasoline-driven centrifugal pump. Figure 2 illustrates the general arrangement of the piping and equipment.

The initial nitrate concentration of raffinate pit 1 water was approximately 320 mg/L. To approximate the required 1.4:1 carbon:nitrogen ratio, 7,000 gal (26,497 L) of the 23% calcium acetate solution were added. To approximate the 50 mg/L concentration in the pit, 220 gal (833 L) of the mono/disodium phosphate was added. The combined nutrient solution was pumped to pit 1 via the recirculation system. Once the prescribed volumes had been delivered, the contents of pit 1 were recirculated for 24 h to ensure good mixing of nutrients. After the mixing was complete, the pit was allowed to remain undisturbed.

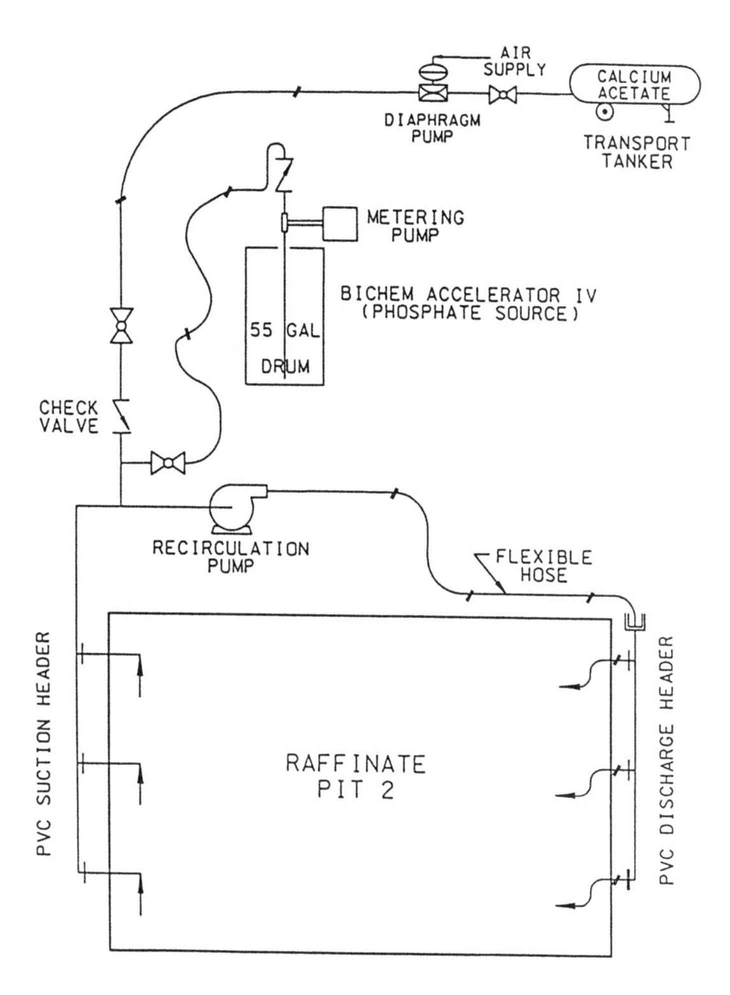

FIGURE 2. General equipment and piping layout at raffinate pit 2.

Temperature, pH, nitrate, DO, total organic carbon (TOC), and sulfate were monitored at each corner of the pond. Figure 3 presents the data generated from the monitoring (only TOC, DO, pH, and nitrate are presented in the graph). These data are consistent with data generated from the previous studies. Approximately 2 weeks into the full-scale test the nitrate removal rate appeared to slow considerably. It was determined that the TOC (calcium acetate) had been effectively depleted. Additional calcium acetate was introduced into pit 1 and was effective at restarting nitrate removal. This carbon depletion/reaction cessation phenomenon was observed in subsequent batches. Each time, addition of calcium acetate remedied the situation, suggesting that theoretical stoichiometric values were not entirely appropriate for the field tests. Adjustments to the TOC dosage are discussed later. The water was transferred for subsequent treatment once the nitrate was below 10 mg/L.

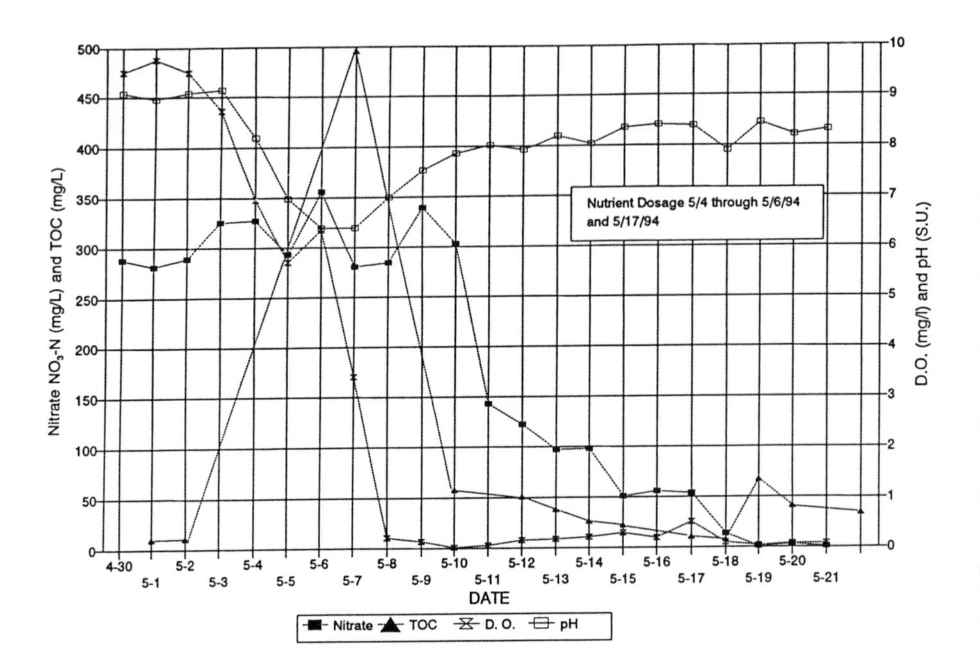

FIGURE 3. Full-scale operation in raffinate pit 1 — batch 1, nitrate reduction trend.

Subsequent full-scale operation used raffinate pits 1 and 2. Both pits can hold approximately 900,000 gal (3,406,860 L). Pit 2 was outfitted with the same equipment as pit 1. Pit 2 was seeded with approximately 30,000 gal (113,462 L) of water from pit 1. This appeared to have significantly shortened the acclimation time required for pit 2 denitrification to occur. Figure 4 is representative of the results of subsequent batches. The time required to denitrify the water decreased as the number of batches increased. This suggested that, once a healthy microbial denitrification population was established, 1-week or less turnaround time per batch could be routinely expected.

The calcium acetate dosing strategy was modified to minimize excess TOC at the completion of each batch. Excess TOC prematurely exhausts the activated carbon system in the subsequent physico/chemical treatment system. Calcium acetate was added at approximately 80% of the theoretical requirement. As the nitrate reduction rate began to decrease, calcium acetate was added in sufficient quantities to complete the reaction to the desired less than 10 mg/L concentration.

DISCUSSION

The microbial diversity required to denitrify these waters apparently has existed for some time. However, due to the nutrient-deficient state in the raffinate pit surface waters, this population did not thrive. When a carbon and phosphate

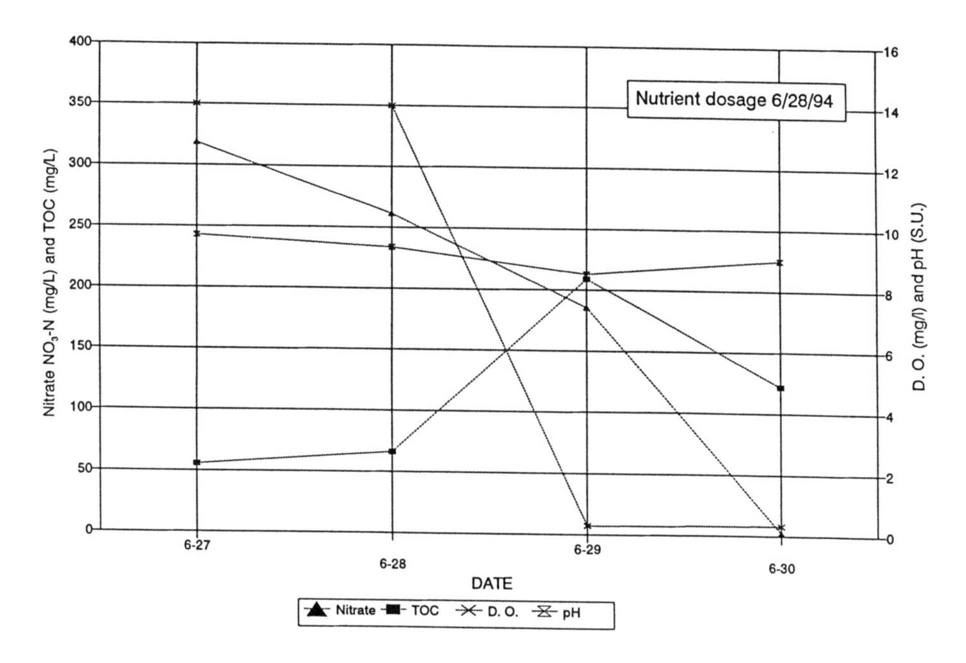

FIGURE 4. Full-scale operation in raffinate pit 2 — batch 5, nitrate reduction trend.

source were provided in sufficient quantities, the indigenous denitrifying populations appeared to flourish.

There was a clear relationship between DO and nitrate mineralization. Each batch illustrated that, when the DO concentration decreased below 2 mg/L, denitrification followed within a short period of time. The effects of wind and rain on the DO concentration were not well-defined in this study. A relationship also exists between available carbon and denitrification. In each case, as the TOC concentration decreased, the denitrification rate also decreased. Then, when TOC (as calcium acetate) was added denitrification increased until the nitrate was effectively depleted. The effect of temperature on full-scale operations was never examined since all the full scale activities occurred during the late spring and early fall. However, bench- and pilot-scale studies suggest that the denitrification rate would significantly decrease as the temperature dropped below 50°F (10°C).

In summary, the WSS had an immediate need to treat nitrate-bearing surface water. The bench- and pilot-scale studies demonstrated that this could be done in a timely and cost effective way (approximately $35/1,000 gal [3,785 L]) via biological denitrification (U.S. DOE 1994b). The pilot studies also provided the necessary engineering data to successfully scale up the operation. The biodenitrification process effectively treated approximately 5.2 million gal (19.7 million L) of surface water from the raffinate pits.

REFERENCES

SITEX Environmental, Inc. 1993. *Report of Initial Study Results for Establishment of Treatment Process for Denitrification of Water in Pit No. 3 Weldon Spring Site Remedial Action Project*, Project No. 7767. November 5, 1993.

U.S. Environmental Protection Agency. 1993. *Nitrogen Control Manual*. EPA/625/R-93/010. Office of Research and Development, Center for Environmental Research Information, and the Office of Wastewater Enforcement and Compliance, Washington, DC.

U.S. Department of Energy. 1994a. *In Situ Treatment Plan for Biodenitrification Field Scale Demonstration*, Rev. 0. DOE/OR/21548-451. Prepared by MK-Ferguson Company and Jacobs Engineering Group, St. Charles, MO.

U.S. Department of Energy. 1994b. *In Situ Biodenitrification of Raffinate Pit Water Pilot Scale Test Report*, Rev. 0. DOE/OR/21548-452. Prepared by MK-Ferguson Company and Jacobs Engineering Group, St. Charles, MO.

Heavy Metal Removal by Caustic-Treated Yeast Immobilized in Alginate

Yongming Lu and Ebtisam Wilkins

ABSTRACT

Saccharomyces cerevisiae yeast biomass was treated with hot alkali to increase its biosorption capacity for heavy metals and then was immobilized in alginate gel. Biosorption capacities for Cu^{2+}, Cd^{2+}, and Zn^{2+} on alginate gel, native yeast, native yeast immobilized in alginate gel, and caustic-treated yeast immobilized in alginate gel were all compared. Immobilized yeasts could be reactivated and reused in a manner similar to the ion exchange resins. Immobilized caustic-treated yeast has high heavy metal biosorption capacity and high metal removal efficiency in a rather wide acidic pH region. The biosorption isotherm of immobilized caustic-treated yeast was studied, and empirical equations were obtained. The initial pH of polluted water affected the metal removal efficiency significantly, and the equilibrium biosorption capacity seemed to be temperature independent at lower initial metal concentrations.

INTRODUCTION

Because conventional physico-chemical treatment methods often are ineffective or uneconomical when the heavy metal concentrations in polluted water are in the range of 10 to 100 mg/L and the permissible concentrations are less than 1 mg/L (Shumate et al. 1978), the search for new and innovative technologies has focused attention on the metal removal capacities of various biological materials, such as bacteria, yeasts, filamentous fungi, algae, and plant cells (Brierley 1991, Pradhan 1992). Biosorption is the accumulation of metals without active uptake. It may occur even when the cell is metabolically inactive (Tsezos 1985). Some methods of killing cells may actually improve biosorption properties of the biomass (Kuyucak and Volesky 1988). Treatment of biomass with NaOH and other alkaline reagents has been demonstrated to increase the capacity of certain biomass to adsorb metals (Brady et al. 1994), and most base-soluble biomass can be reconstituted by adding acid to attain neutral pH (Brierley and Brierley 1993).

Immobilization of biomass in a polymeric matrix can yield beads or granules with optimum size, mechanical strength, rigidity, and porosity characteristics. This allows the metals to be stripped and the biomass to be reactivated and reused in a manner similar to ion exchange resins. Because of the presence of some functional groups for binding divalent cations, calcium alginate, one of the popular entrapment materials used in immobilization, has been extensively investigated as a biopolymer for binding heavy metals from diluted aqueous solutions (Jang et al. 1990a, 1990b).

Because of the ready availability of waste *Saccharomyces cerevisiae* yeast biomass from classical food fermentation industries, the aim of this study is to investigate the metal biosorption capacity of caustic-treated yeast entrapped in alginate gel. The biosorption capacities of alginate gel, native yeast, and immobilized yeast without alkali treatment also were studied.

MATERIALS AND METHODS

Preparation of Biosorbent Beads

Yeast biomass, *Saccharomyces cerevisiae* (bakers' yeast, type II), and alginic acid [sodium salt with high viscosity, from *Macrocystis pyrifera* (kelp)] were obtained from Sigma Chemical Co.

Native yeast (6 g) was mixed with 20 mL 0.75 M NaOH, and the resulting solution was heated to 70 to 90°C for 10 to 15 min. After the temperature of the solution returned to ambient, 4 M HCl was used to adjust the pH to 6 to reconstitute the base-soluble biomass. The product was then mixed with 30 mL 3.33% (w/w) sodium alginate solution. The immobilized caustic-treated yeast (ICY) was prepared by forcing the mixture through a 26-gauge needle into $CaCl_2$ (1.5% w/v) solution. The immobilized yeast (IY) was prepared with the same recipe (12% yeast and 2% alginate) without the alkali treatment process. The Ca-alginate gel beads were prepared by forcing the sodium alginate (4% w/w) through a 26-gauge syringe (2-mm ID) into $CaCl_2$ (1.5% w/v) solution. All three types of biosorbent beads were cured overnight in the $CaCl_2$ solution, rinsed with deionized water (DW) 3 times for about 3 min, soaked in 0.5 M HCl solution for more than 24 h, rinsed with deionized water 3 times for about 3 min again, soaked in deionized water for more than 24 h, and then rinsed with deionized water before the metal adsorption capacity studies. All beads (alginate gel bead, immobilized yeast, and immobilized caustic treated yeast) prepared in this study had diameters of 2.3 ± 0.2 mm.

Biosorption Capacity

Experiment. Preliminary experiments were performed to determine steady-state equilibrium time in a 400-mL beaker for native yeast, alginate gel bead, immobilized yeast, and immobilized caustic-treated yeast. 0.4 g of beads dry mass was added to 200 mL well mixed metal solution of either Cu^{2+}, Cd^{2+}, or Zn^{2+}

with initial pH 5.0 at ambient temperature (23 ± 1°C). Samples were periodically withdrawn and assayed for metal content. The equilibrium times found were used for other experiments. The metal biosorption capacities of native yeast, alginate gel bead, immobilized yeast, and immobilized caustic-treated yeast were studied by adding 0.2 g (dry mass) of beads or native yeast to 100 mL of metal solution of either Cu^{2+}, Cd^{2+}, or Zn^{2+} with initial pH 5.0, and the initial concentration was in the range from 16 to 18 mg/L at ambient temperature (23 ± 1°C). For native yeast, samples were separated in a high speed centrifuge before the metal ion content assay. The effect of temperature on the biosorption capacity was studied by keeping batch reactors at 7 ± 1°C, 23 ± 1°C, or 45 ± 1°C. The effect of initial pH on the biosorption capacity was studied by adding HCl or NaOH to adjust the initial pH of the solution. The biosorption isotherm was studied by using different concentrations of the metals ranging from 2 to 1,000 mg/L with fixed biosorbent density (2 g/L), or by using different biosorbent density (2 to 6 g/L) and fixed initial metal concentration (16 to 18 mg/L). The initial pH was 5.0, and the experiments were completed at ambient temperature (23 ± 1°C).

Calculation. Metal-laden gel beads, collected after equilibrium was reached in batch experiments, were dried at about 110°C for 2 h and weighed. The biosorption capacity was determined by the difference between the initial metal ion concentration and the final one after equilibrium was reached. Metal concentrations were quantified using a Perkin-Elmer atomic absorption spectrophotometer Model 460. For comparison, the digested biomass was also used to determine the metal biosorption capacity. The procedure was as follows: 5 mL concentrated HNO_3 (70%) and 10 mL deionized water were added to the dried beads in a 125-mL Erlenmeyer flask, and the mixture was heated at boiling point for about 15 min. The mixture was centrifuged, and the solution was diluted to a fixed volume using deionized water for metal content assay.

Metal Desorption and Bead Reuse. Two tenths g (dry mass) of immobilized yeast (or immobilized caustic treated yeast) was added to 100 mL of metal solution of either Cu^{2+}, Cd^{2+}, or Zn^{2+} with an initial pH of 5.0; and the initial concentration was in the range from 16 to 18 mg/L at ambient temperature (23 ± 1°C). Following initial biosorption of metals, the metal-laden beads were soaked in 25 mL 0.5 M HCl for more than 24 h for desorption of metals, rinsed with deionized water several times, soaked in deionized water for more than 24 h, and then rinsed with deionized water before the next biosorption cycle. The initial metal concentration and that after equilibrium were assayed for every cycle, and the biosorbent beads were dried and weighed after the seventh cycle study.

RESULTS AND DISCUSSION

Preliminary experiments were carried out to determine the equilibrium time for biosorption. About 36 h were used as equilibrium time for immobilized yeast, immobilized caustic treated yeast, or Ca-alginate bead; 2 h were used as equilibrium time for native yeast. The metal biosorption capacities of alginate gel, native yeast, immobilized yeast, and immobilized caustic treated yeast were compared. The average results are listed in Table 1. Alginate gel had the highest metal binding capacity compared to other types of biosorbent. Immobilized yeast reduced the quantity of heavy metal binding to the biomass as compared to biosorption by native yeast by about 10 to 25%. This decrease of metal binding capacity probably occurred because of crosslinking of potential metal-binding sites with alginate gel and masking of active sites due to the higher density of the biomass. However, yeast treated with hot NaOH obviously had increased metal binding capacity by about 20%, compared to biosorption by native yeast. On the other hand, the increase of biosorption capacity due to the presence of alginate with highest metal biosorption capacity in immobilized caustic-treated yeast was insignificant, because alginate is only 14% of the total weight.

For immobilized biomass, it is possible to regenerate the biosorbent and recover the loaded metals after desorption in a manner similar to ion exchange resins. HCl was used as a desorbent in this study because of the availability and the low cost. The results for Cu^{2+}, Cd^{2+}, and Zn^{2+} biosorbed by immobilized yeast and immobilized caustic-treated yeast are shown in Figure 1 for 7 cycles. No decrease in biosorption capacity and no biomass loss were found in the experimental period.

The effect of metal concentration after equilibrium on the metal biosorption capacity can be described by the Langmuir and Freundlich adsorption isotherms. For the immobilized caustic-treated yeast system, both isotherms fit the data well for Cu^{2+} and Cd^{2+} biosorption. But for Zn^{2+}, the Freundlich isotherm fits the data better than the Langmuir isotherm. It is impossible for biosorption capacity to increase with equilibrium concentration exponentially at high concentrations, because biosorption saturation is physically reasonable. For the Freundlich isotherm, using two different sets of model parameters for different concentration regions gives a better fit than using one set of parameters only. The concentration at which model parameters changed was about 1×10^{-4} M (10, 4, and 6 mg/L for Cd^{2+}, Cu^{2+}, and Zn^{2+}, respectively). The series of decreasing sorption is $Cu^{2+} > Cd^{2+} > Zn^{2+}$. The experimental equations obtained through the least square

TABLE 1. Metal biosorption capacity (mg/g dry mass) of different types of biosorbents with initial concentrations 17 ± 1 mg/L and pH 5.0 at 23 ± 1°C. (IY-immobilized yeast, ICY-immobilized caustic-treated yeast).

Species	Alginate Gel	Native Yeast	IY	ICY
Cu^{2+}	9.01	7.02	6.32	8.46
Cd^{2+}	7.50	5.85	4.44	6.20
Zn^{2+}	7.08	4.54	3.73	5.72

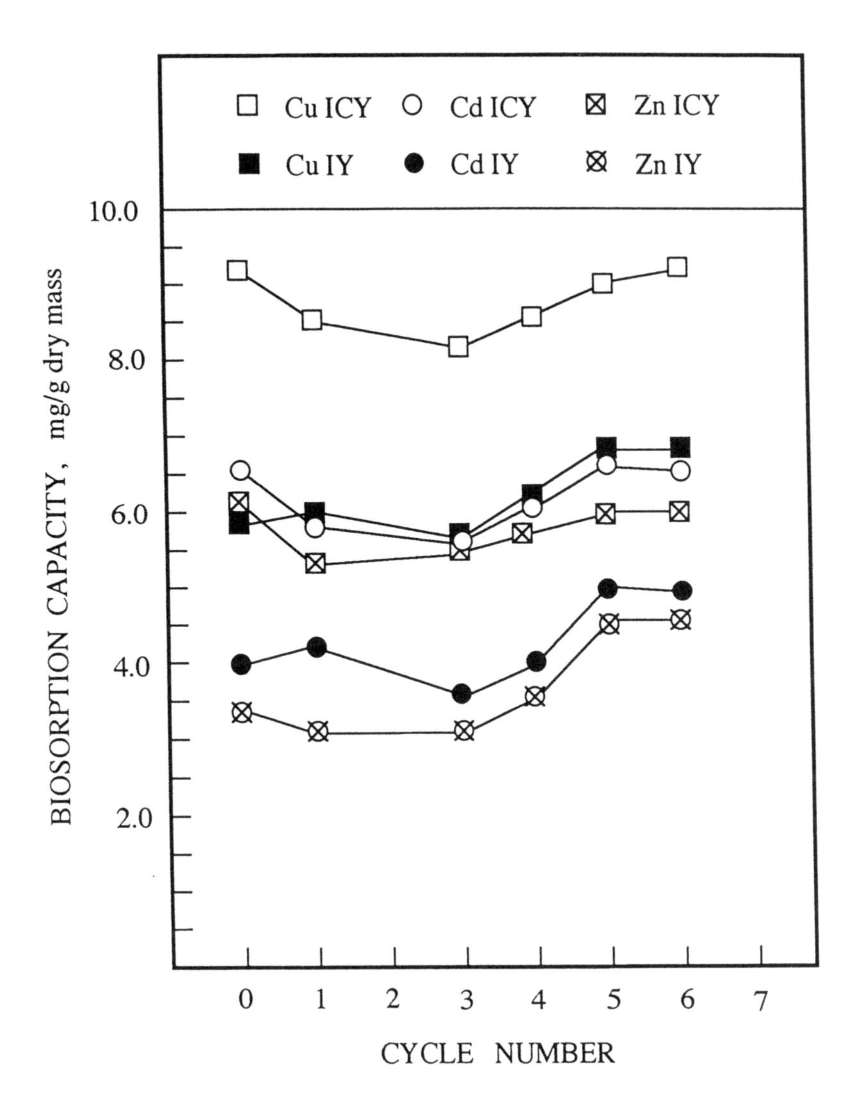

FIGURE 1. Biosorption of metal ions to immobilized yeast (IY) and immobilized caustic-treated yeast (ICY) reactivated and reused as ion exchange resins with initial concentrations in the range of 16 to 18 mg/L and pH 5.0 at 23 ± 1°C.

method are listed below, where Q is biosorption capacity (mg/g dry mass) and C is the metal concentration after equilibrium (mg/L).

Langmuir isotherm

$$1/Q = 0.285/C + 0.0286 \quad \text{for } Cu^{2+} \qquad (1)$$

$$1/Q = 0.681/C + 0.0413 \quad \text{for } Cd^{2+} \qquad (2)$$

$$1/Q = 0.401/C + 0.114 \quad \text{for } Zn^{2+} \qquad (3)$$

Freundlich isotherm

$$\log Q = 0.943 \log C + 0.500 \quad \text{for } Cu^{2+} \quad (C < 4 \text{ mg/L}) \tag{4}$$
$$\log Q = 0.250 \log C + 0.900 \quad \text{for } Cu^{2+} \quad (C > 4 \text{ mg/L}) \tag{5}$$
$$\log Q = 0.743 \log C + 0.193 \quad \text{for } Cd^{2+} \quad (C < 10 \text{ mg/L}) \tag{6}$$
$$\log Q = 0.343 \log C + 0.593 \quad \text{for } Cd^{2+} \quad (C > 10 \text{ mg/L}) \tag{7}$$
$$\log Q = 0.536 \log C + 0.268 \quad \text{for } Zn^{2+} \quad (C < 6 \text{ mg/L}) \tag{8}$$
$$\log Q = 0.275 \log C + 0.475 \quad \text{for } Zn^{2+} \quad (C > 6 \text{ mg/L}) \tag{9}$$

Metal removal was significantly affected by the initial pH of the metal solution due to cation competition effects with H^+ ion. Biosorption of Cu^{2+}, Cd^{2+}, and Zn^{2+} on immobilized caustic-treated yeast did not occur below pH 3, but increased rapidly above pH 3, and leveled off at pH 4 to 6 (Figure 2). The second

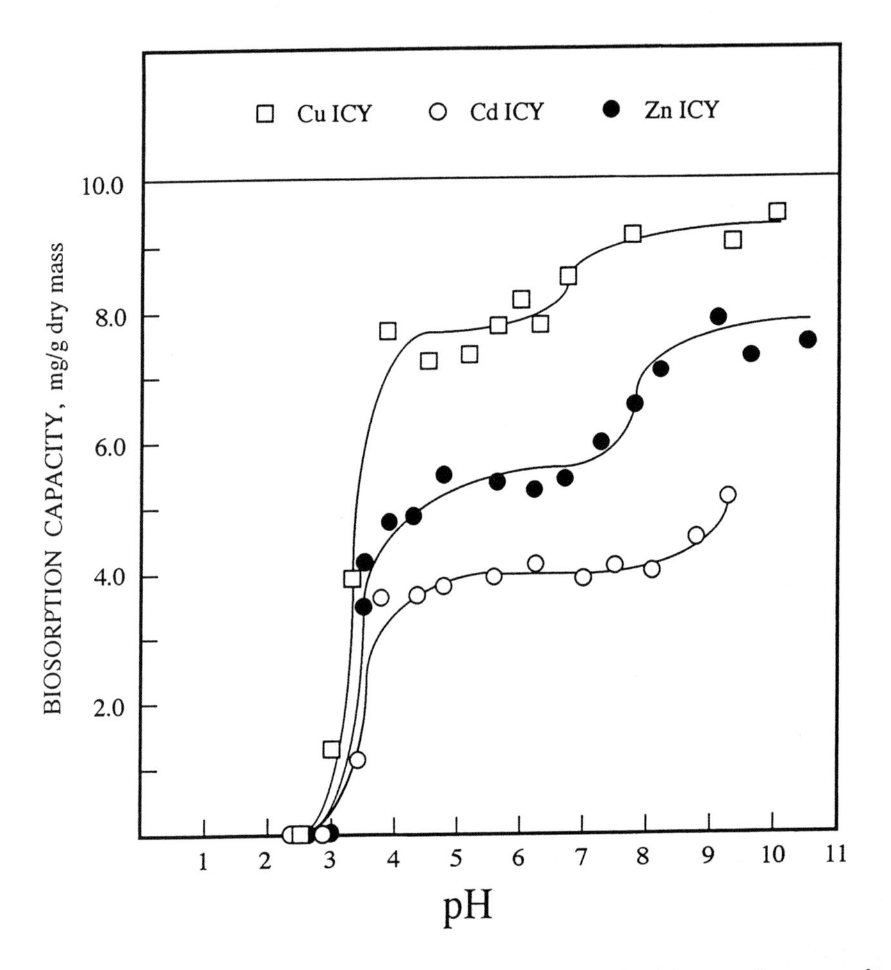

FIGURE 2. Effect of initial pH of metal solution on the biosorption capacity of immobilized caustic-treated yeast (ICY) with initial concentration of 20, 13, and 18 mg/L for Cu^{2+}, Cd^{2+}, and Zn^{2+}, respectively, at 23 ± 1°C.

TABLE 2. Effect of temperature on the metal biosorption capacity (mg/g dry mass) of immobilized caustic-treated yeast (ICY) with initial concentrations of 18, 14, and 18 mg/L for Cu^{2+}, Cd^{2+}, and Zn^{2+}, respectively, and pH 5.0 at 23 ± 1°C.

	Temperature (°C)			Ave. Equil.	Calculated Values	
Species	7 ± 1°C	23 ± 1°C	45 ± 1°C	Conc. (mg/L)	Langmuir	Freundlich
Cu ICY	6.45	6.69	7.52	2.5	7.01	7.50
Cd ICY	4.44	4.58	4.25	4.3	5.01	4.61
Zn ICY	5.24	5.53	5.41	5.9	5.50	4.80

significant increase of biosorption capacity occurred at pH 6, 7, and 9 for Cu^{2+}, Zn^{2+}, and Cd^{2+}, respectively. This may demonstrate the presence of chemical precipitation due to hydrolysis of metal ions. The order of pH at which the second increase of biosorption capacity occurred is the same as that of the solubility products of metal hydroxides, $Cu(OH)_2$ ($K_{sp}=10^{-19.8}$) $<Zn(OH)_2$ ($K_{sp} = 10^{-16.8}$) $<Cd(OH)_2$ ($K_{sp} = 10^{-13.9}$) (Schwitzgebel and Manis 1994). Because the concentration of Cd^{2+} after equilibrium was lower than that of Zn^{2+}, due to the lower initial concentration of Cd^{2+} (14 mg/L) compared to that of Zn^{2+} (18 mg/L), the biosorption capacity of Zn^{2+} at the leveled pH region was higher than that of Cd^{2+}.

The effect of temperature on the metal biosorption capacity is shown in Table 2. The results demonstrate that the biosorption is temperature independent. The data calculated using the equations (1, 2, 3, 4, 6, and 8) are listed in Table 2 and show the validity of both isotherms.

CONCLUSIONS

Treating yeast treated with hot alkali enhanced its heavy metal biosorption capacity significantly. Immobilized yeasts could be reactivated and reused in a manner similar to ion exchange resins. The biosorption could be described by the Langmuir and Freundlich adsorption isotherms and was temperature independent at lower initial metal concentration. On the other hand, the initial pH of heavy metal solution significantly affected the metal removal efficiency.

REFERENCES

Brady, D., A. Stoll, and J. R. Duncan. 1994. "Biosorption of heavy metal cations by non-viable yeast biomass." *Environ. Technol. 15*: 429-428.

Brierley, C. L., and J. A. Brierley. 1993. "Immobilization of biomass for industrial application of biosorption." In Torma, A. E., Apel, M. L., and Brierley, C. L. (Eds.), *Biohydrometallurgical*

Technologies, Vol. 2, pp. 35-44. Proceedings of an international biohydrometallurgy symposium, Wyoming, USA, Aug. 22-25, 1993. The Minerals, Metals and Materials Society.

Brierley, C. L. 1991. "Bioremediation of metal-contaminated surface and ground waters." *Geomicrobiol. J. 8*: 201-223.

Jang, L. K., W. Brand, M. Resong, W. Mainieri, and G. G. Geesey. 1990a. "Feasibility of using alginate to absorb dissolved copper from aqueous media." *Environ. Prog. 9*: 269-274.

Jang, L. K., G. G. Geesey, S. L. Lopez, S. L. Eastman, and P. L. Wichlacz. 1990b. "Use of a gel forming biopolymer directly dispensed into a loop fluidized bed reactor to recover dissolved copper." *Water Res. 24*: 889-897.

Kuyucak, N., and B. Volesky. 1988. "Biosorbents for recovery of metals from industrial solutions." *Biotechnol. Lett. 10*(2): 137-142.

Pradhan, A. A., and A. D. Levine. 1992. "Experimental evaluation of microbial metal uptake by individual components of a microbial biosorption system." *Water. Sci. Technol. 26*(9-11): 2145-2148.

Schwitzgebel, K., and D. M. Manis. 1994. "Removal of chromate, cyanide, and heavy metals from wastewater." In D. L. Wise and D. J. Trantolo (Eds.), *Process Engineering for Pollution Control and Waste Minimization*, pp. 535-556. Marcel Dekker Inc., New York, NY.

Shumate, S. E. II., G. W. Strandberg, and J. R. Parrott, Jr. 1978. "Biological removal of metal ions from aqueous process streams." *Biotechnol. Bioeng. Symp. 8*: 13-20.

Tsezos, M. 1985. "The selective extraction of metals from solution by microorganism: A brief overview." *Can. Metall. Q.* 24(2): 141-144.

Mechanisms Regulating the Reduction of Selenite by Aerobic Gram (+) and (–) Bacteria

Carlos Garbisu, Takahisa Ishii, Nancy R. Smith, Boihon C. Yee, Don E. Carlson, Andrew Yee, Bob B. Buchanan, and Terrance Leighton

ABSTRACT

Toxic species of selenium are pollutants found in agricultural and oil refinery wastestreams. Selenium contamination is particularly problematic in areas that have seleniferous subsurface geology, such as the central valley of California. We are developing a bacterial treatment system to mitigate selenium-contaminated wastestreams using *Bacillus subtilis* and *Pseudomonas fluorescens*, respectively, as model gram (+) and (–) soil bacteria. We have found that, during growth, both organisms reduce selenite, a major soluble toxic species, to red elemental selenium — an insoluble product generally regarded as nontoxic. In both cases, reduction depended on growth substrate and was effected by an inducible system that effectively removed selenite at concentrations typical of polluted sites — i.e., 50 to 300 μg/L. The bacteria studied differed in one respect: when grown in medium supplemented with nitrate or sulfate, the ability of *P. fluorescens* to remediate selenite was enhanced, whereas that of *B. subtilis* was unchanged. Current efforts are being directed toward understanding the biochemical mechanism(s) of detoxification and determining whether bacteria occurring in polluted environments such as soils and sludge systems are capable of selenite remediation.

INTRODUCTION

Selenium pollution is the most widely encountered example of group V and VI metal hazards (Burton et al. 1987, Heinz et al. 1990). Toxic species of selenium are pollutants found in groundwater, smelting effluents, agricultural and municipal wastewater, waste disposal sites, power plant cooling reservoirs, and oil refinery wastestreams. In the best known case of selenium pollution — agricultural drainage water that had percolated through seleniferous soils and was discharged into the Kesterson National Wildlife Refuge — selenium accumulated to 100 times the levels found in normal surface water (Heinz et al. 1990). Wildlife deformities and deaths from toxic selenium species resulted in the closure of Kesterson in 1986.

A number of microorganisms can reduce selenite, a soluble toxic form, to a red insoluble product (elemental selenium), which is considered nontoxic. Organisms accomplishing this reaction in the soil, primarily mixed populations of bacilli, have been described for sites rich in selenite (Burton et al. 1987). Selenite has, on the one hand, been found to be used as an electron acceptor in an anaerobic form of respiration (Macy et al. 1989) and, on the other, to be reduced independently of dissimilatory electron transport (Lortie et al. 1992). In the latter case, the authors concluded that reduction was used as an aerobic mechanism of selenite detoxification.

Despite reports describing its occurrence, the physiological mechanisms responsible for the detoxification of selenite have not been elucidated for aerobic bacteria. We have examined the physiological regulation of selenite reduction to elemental selenium by well-characterized laboratory strains of representative gram (+) and (−) aerobic soil bacteria, typified, respectively, by *Bacillus subtilis* and *Pseudomonas fluorescens*.

EXPERIMENTAL PROCEDURES AND MATERIALS

The organisms were cultured on a minimal chemically defined liquid medium (the MOPS medium) (Pierce et al. 1992). Cultures were grown in 250 mL Erlenmeyer flasks with continuous shaking (250 rpm) at 37 and 30°C for *B. subtilis* and *P. fluorescens*, respectively. Before initiating experiments, the cultures were transferred twice into fresh medium, over a period of approximately 24 h. Reinoculations were timed to ensure that cultures were under excess nutrient conditions, and thus in an environment allowing balanced exponential growth. Growth experiments were repeated a minimum of three times with consistent results. Data from representative experiments are presented.

Growth was routinely monitored by measuring absorbance at 600 nm in a Beckman/Gilford spectrophotometer. Measurements based on viable cell counts confirmed that absorbance at 600 nm accurately reflects growth.

Selenium was determined with the supernatant fraction obtained by centrifuging samples at 15,000 *x* g for 10 min at 4°C. Samples were analyzed by inductively coupled plasma spectrometry using a Perkin-Elmer Plasma 40 emission spectrophotometer.

RESULTS

Our first objective was to determine whether both bacterial species were capable of growth in the presence of selenite. To test this possibility, cells grown in a chemically defined medium were used as inoculum for this same medium supplemented with sodium selenite (1 mM or 79 mg/L selenium). Based on absorbance measurements, the *B. subtilis* and *P. fluorescens* cultures reached maximal growth after 30 and 48 hr, respectively (Figure 1). The cultures turned red as growth progressed, indicating that selenite was being converted to a form that was earlier identified as red elemental selenium (Buchanan et al. 1995). Although a significant amount of the selenite was removed from the supernatant in both cases, *B. subtilis* appeared somewhat more efficient in reducing selenite than *P. fluorescens*. Similar levels of growth were observed as determined by total viable counts (data not shown).

When *B. subtilis* cultures were supplemented with concentrations of selenite representative of those occurring in polluted sites (i.e., from 50 to 300 μg/L

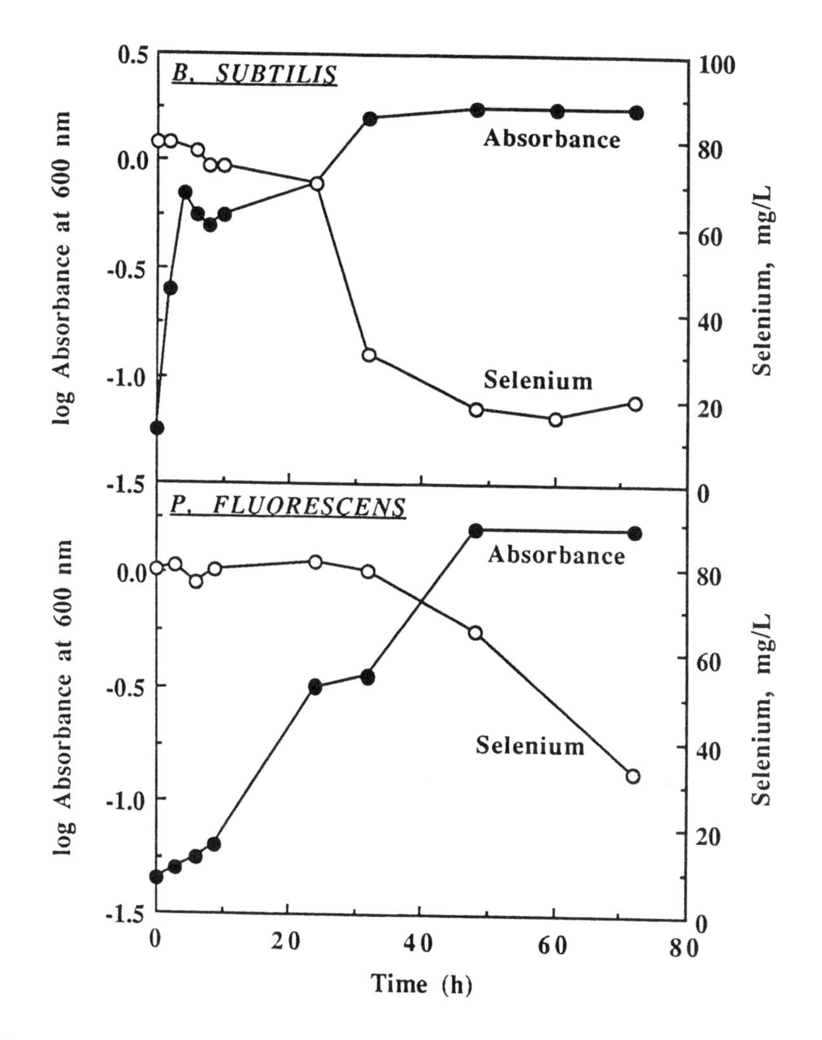

FIGURE 1. Effect of 1 mM sodium selenite (79 mg/L selenium) on *B. subtilis* and *P. fluorescens* growth, and the solubility status of selenium.

selenium), the cells demonstrated the previously observed pattern of reduction. With 50 µg/L, the selenium level in the supernatant fraction decreased to less than 3 µg/L after 10 hr of incubation — i.e., 95% of the added selenite was removed. When minimal medium containing 1 mM selenite was inoculated with *B. subtilis* or *P. fluorescens* cells previously grown in the same selenite medium, the culture (designated "selenite-induced") grew rapidly and, in the case of *B. subtilis,* did not show the decrease in optical density consistently found after 4 to 5 h of incubation with uninduced cultures (Figure 2 vs. Figure 1). Similarly,

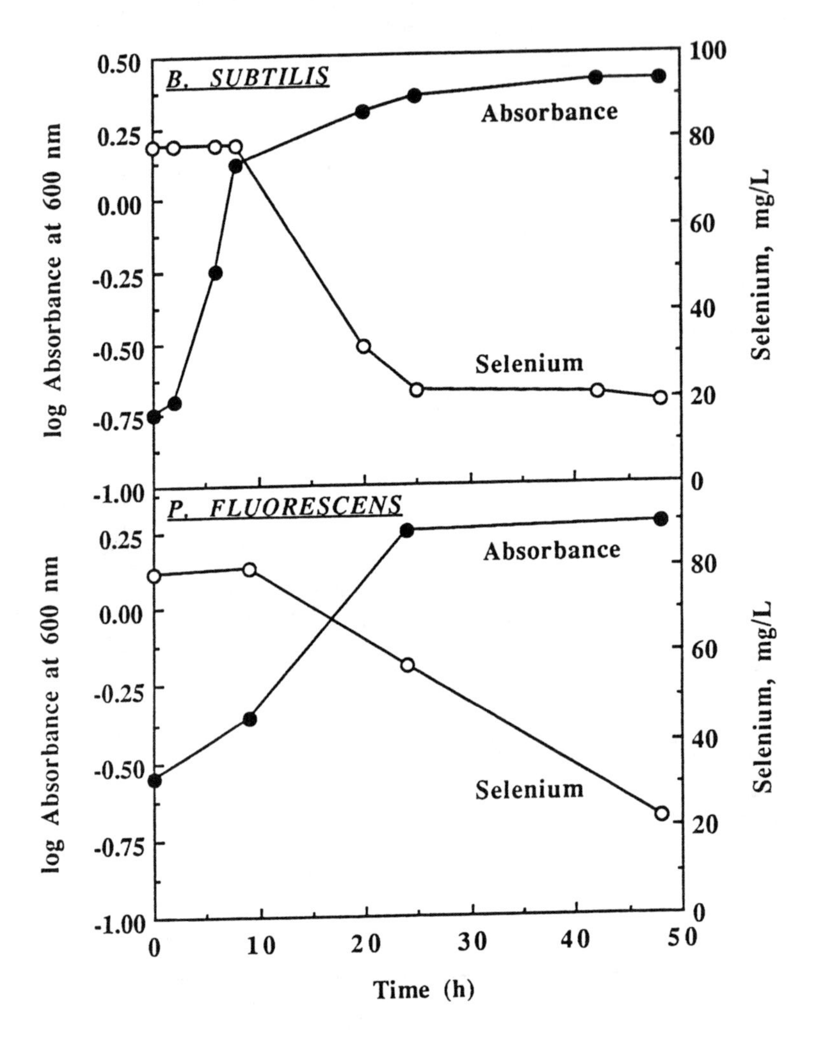

FIGURE 2. Growth and solubility status of selenium in "induced" *B. subtilis* and *P. fluorescens* cultures supplemented with 1 mM sodium selenite (inoculum was taken from a culture grown in the presence of 1 mM sodium selenite).

the increase in growth rate was reflected in the finding that the induced organisms reached stationary phase earlier than their noninduced counterparts. In addition, although the final selenium concentration in the supernatant fraction was similar in both the noninduced (Figure 1) and induced (Figure 2) *B. subtilis* cultures (around 20 mg/L), the latter appeared to remediate selenite at a faster rate — i.e., the selenium level in the supernatant fraction decreased from 79 mg/L to approximately 20 mg/L after 24 rather than 48 hr. The induced *P. fluorescens* culture showed an enhancement in both the rate and (unlike *B. subtilis*) the extent of selenite reduction (Figure 1 vs. Figure 2).

In the experiments described thus far, cells were grown with glucose as carbon source. As the bacteria studied are known to use carbon sources other than glucose, it was of interest to test several substrates for their ability to support selenite bioremediation, namely, sucrose, citrate, and glycerol. The concentrations of the substrates used (35 mM sucrose, 140 mM glycerol and 70 mM sodium citrate) were chosen to equalize the number of carbon atoms provided in the culture medium. The growth rates and final cell densities were similar with all substrates tested (data not shown). However, the capacity for supporting selenite reduction varied: sucrose and glycerol supported remediation, whereas citrate did not (Figure 3).

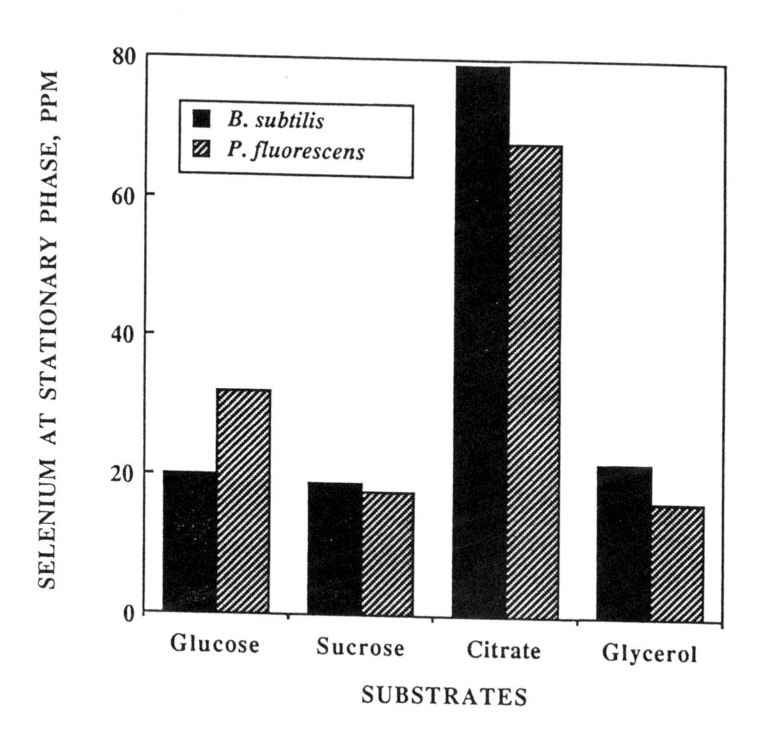

FIGURE 3. Effect of carbon source on *B. subtilis* and *P. fluorescens* remediation of selenite. Cultures were supplemented with 1 mM sodium selenite (79 mg/L selenium) and grown to stationary phase.

One of the features characterizing anaerobic selenite remediation systems is a sensitivity to nitrate and sulfate, both of which are preferred as anaerobic electron acceptors. Our results show that aerobic bacteria differ in this regard. The remediation of selenite by *B. subtilis* was independent of nitrate and sulfate, whereas that of *P. fluorescens* was enhanced (Figure 4).

DISCUSSION

The present results demonstrate that well-characterized laboratory strains of common soil bacteria, *B. subtilis* and *P. fluorescens*, remove selenite from solution

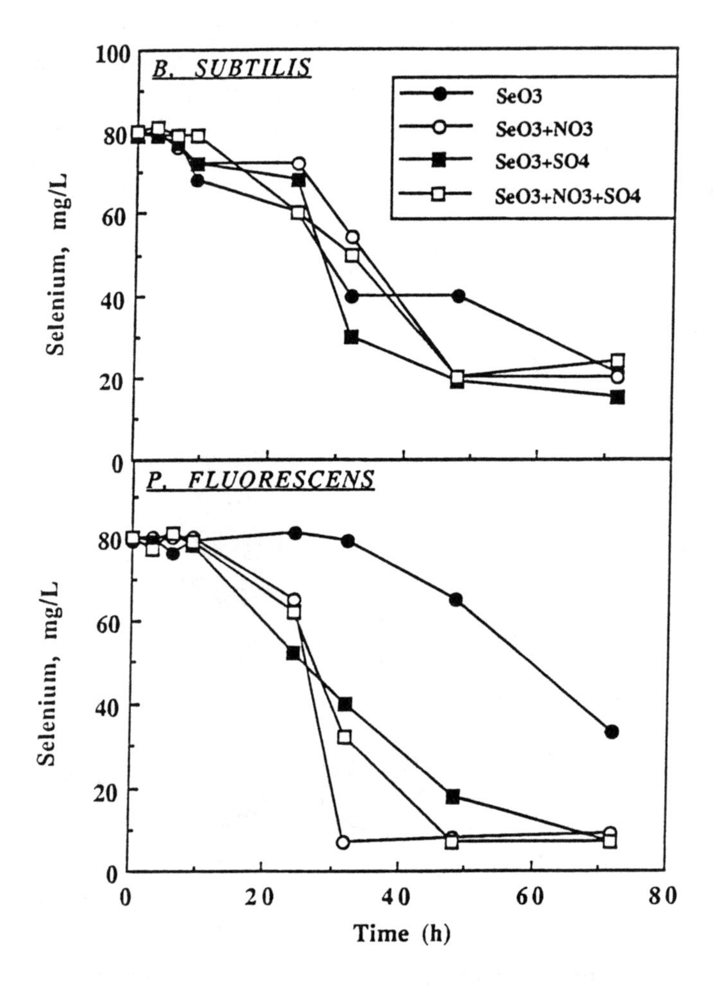

FIGURE 4. Effect of 10 mM sodium nitrate and/or sodium sulfate on *B. subtilis* and *P. fluorescens* remediation of selenite. Cultures were supplemented with 1 mM sodium selenite (79 mg/L selenium).

by reduction to a product earlier identified as a red form of elemental selenium. The finding that both species can quantitatively remediate selenite in the 50 to 300 ng/mL range raises the possibility that organisms of this type may be useful for environmental restoration and liquid waste treatment.

The *B. subtilis* and *P. fluorescens* system(s) of selenite reduction is inducible and dependent on the carbon substrate provided for growth. The finding that the remediation of selenite by *B. subtilis* was independent of nitrate and sulfate suggests that selenite is not reduced via dissimilatory electron transport but via a novel detoxification system. Additional studies are required to explain the enhancing effect of nitrate and sulfate on selenite reduction by *P. fluorescens*. Experiments are in progress to identify the mechanism responsible for reducting selenite in both organisms. In addition, we are currently assessing the capability of bacterial populations residing in soils and sludge systems to remediate selenite.

REFERENCES

Buchanan, B. B., J. J. Bucher, D. E. Carlson, N. M. Edelstein, E. A. Hudson, N. Kaltsoyannis, T. Leighton, W. Lukens, H. Nitsche, T. Reich, K. Roberts, D. K. Shuh, P. Torretto, J. Woicik, W.-S. Yang, A. Yee, and B. C. Yee. 1995. "A XANES and EXAFS Investigation of the Speciation of Selenite Following Bacterial Metabolization." *Inorganic Chemistry*. In press.

Burton, G. A., Jr., T. H. Giddings, P. DeBrine, and R. Fall. 1987. "High Incidence of Selenite Resistant Bacteria from a Site Polluted with Selenium." *Applied and Environmental Microbiology*. *53*: 185-188.

Heinz, G. H., G. W. Pendleton, A. J. Krynitsky, and L. G. Gold. 1990. "Selenium Accumulation and Elimination in Mallards." *Archives of Environmental Contamination and Toxicology 19*(3): 374-379.

Lortie, L., W. D. Gould, S. Rajan, R. G. L. McCready, and K.-J. Cheng. 1992. "Reduction of Selenate and Selenite to Elemental Selenium by a *Pseudomonas stutzeri* Isolate." *Applied and Environmental Microbiology 58*: 4042-4044.

Macy, J. M., T. A. Michel, and D. G. Kirsch. 1989. "Selenate Reduction by a *Pseudomonas* Species: a New Mode of Anaerobic Respiration." *FEMS Microbiology Letters 61*: 195-198.

Pierce, J. A., C. Robertson, and T. Leighton. 1992. "Physiological and Genetic Strategies for Enhanced Subtilisin Production by *Bacillus subtilis*." *Biotechnology Progress 8*: 211-218.

On-Site Biological Nitrogen Removal Using Recirculating Trickling Filters

James W. Graydon, Stewart M. Oakley, Brian H. Reed, and Harold L. Ball

ABSTRACT

Nitrogen discharged from on-site wastewater treatment and disposal systems often is a source of nitrate contamination in groundwater. Research directed at improving on-site nitrogen removal has led to the development of design criteria and performance data for septic tanks retrofitted with a recirculating trickling filter (RTF) system. The RTF system is based on the process of biological nitrogen removal by sequential aerobic nitrification ($NH_3 \rightarrow NO_3^-$) and anaerobic denitrification ($NO_3^- \rightarrow N_2$). Septic tank effluent is oxygenated by recirculating at 10 to 20 L/min through a 61-cm-diameter by 91-cm-deep trickling filter consisting of a synthetic fixed-growth medium. The oxygenated (nitrified) effluent then returns to the inlet side of the septic tank where high biochemical oxygen demand (BOD) levels and anaerobic conditions promote denitrification. Preliminary data from three RTF systems retrofitted to existing septic tanks show an average effluent total N of 12 ± 7 mg/L and an average N removal of $66 \pm 29\%$. This removal compares favorably with N removal and effluent quality data from intermittent sand filters, recirculating gravel filters, and buried sand filter/greywater (RUCK) systems.

INTRODUCTION

Conventional on-site wastewater treatment and disposal systems (OWTDS), consisting of a septic tank followed by a subsurface drain field, are often a source of nitrate to groundwater. OWTDS are used by approximately one-third of the homes in the United States (Canter and Knox 1985). These systems discharge roughly 1.4×10^{10} L of domestic wastewater containing 5.6×10^5 kg of nitrogen to the soil each day. Typically, OWTDS are designed to reduce the levels of BOD and suspended solids and maximize the hydraulic capacity of the drain field. Nitrogen removal has been neither an objective nor a requirement of conventional OWTDS. Given the magnitude of nitrate impacts to groundwater and the cost of centralized sewer systems, economical processes for on-site nitrogen removal are an important area of wastewater treatment research (Lamb et al. 1990).

Nitrogen Dynamics in Conventional OWTDS

Untreated domestic wastewater typically contains 40 to 50 mg/L total N with the majority occurring as organic nitrogen compounds. In a conventional septic tank, organic-N is reduced to NH_3 (and, at near-neutral pH, NH_3 is protonated to form NH_4^+) such that septic tank effluent typically contains 7 to 21 mg/L organic-N and 28 to 34 mg/L NH_4^+ (Canter and Knox 1985). In a well-designed conventional drainfield, diffusion of O_2 in the subsurface promotes the oxidation of NH_4^+, a relatively immobile cation, to NO_3^- (nitrification) within several inches to several feet below the leachline (Wilhelm et al. 1994). Depending on soil moisture content and the concentration and bioavailability of residual organics, conditions deeper in the soil column may promote the reduction of NO_3^- to N_2O and N_2 (denitrification). Investigations of nitrate levels in soil moisture below drain fields in the community of Los Osos, California have shown nitrate reductions of 42 to 85% that may be attributable to denitrification (San Luis Obispo County 1994; Metcalf & Eddy 1995).

OWTDS Incorporating Nitrogen Removal

The sequential aerobic/anaerobic process described above is the basis of all biological nitrogen removal (BNR) technologies used in wastewater treatment. Several types of advanced OWTDS that remove up to 60% of the total N in domestic wastewater have been developed (Laak et al. 1981; Lamb et al. 1991; Nolte 1992; Cagle and Johnson 1994).

Another approach to on-site BNR using RTF was recently studied in Oregon (Ball 1994). Eight months of operational data from the Oregon RTF system showed average effluent total N concentrations of 16 to 20 mg/L representing an N removal of 71 to 78% compared to the septic tank effluent before the RTF was installed. Based on these data, a pilot testing program for RTF systems was proposed for Butte County, California, to evaluate alternatives to a centralized sewer collection system.

METHODS AND MATERIALS

The overall purpose of this study was to (1) select and retrofit six existing septic tanks with RTFs; (2) to systematically monitor the systems over a period of 6 months; and (3) to develop design, monitoring, operations, and maintenance criteria for RTF (retrofit) systems. Six RTF systems were installed on existing septic tanks (three single-family, two multifamily, and one commercial system) during the summer of 1994. Due to undersized septic tanks, hydraulic overloading of both multifamily systems (Sites 2 and 6) precluded the collection of meaningful data.

The design of the RTF system as retrofitted on existing two-compartment concrete septic tanks is illustrated in Figure 1. Accu-Pac CF-1200 (Brentwood Industries) was used as the trickling filter medium. A 61-cm-deep filter was installed at Site 1 and 91-cm-deep filters were installed on all other systems, giving packing volumes of 180 L and 270 L, respectively. Because Accu-Pac CF-1200 has a surface area of 69 m^2/m^3, it is estimated that the BOD loading rate was, at startup, approximately 0.80 $kg/m^3/day$ for the 61-cm filter and 0.48 $kg/m^3/day$ for the 91-cm filter. It has been reported that organic loading rates in the range of 0.19 to 0.29 $kg/m^3/day$ are necessary for nitrification to occur in trickling filters (Metcalf and Eddy 1991), thus it can be expected that the system will not nitrify until BOD loadings are lowered during the acclimation period.

After a period of acclimation of several weeks to several months, septic tank effluent was monitored periodically for total Kjeldahl nitrogen (TKN), NH_3-N, NO_3^--N, and BOD_5. Samples were collected, preserved, and analyzed in accordance with procedures specified by Greenberg et al. 1985. Specifically, NH_3-N and NO_3-N were measured using ion selective electrodes and the micro-Kjeldahl method was used for TKN (Section 4500); BOD_5 was measured according to Section 5210B. In addition, all samples were analyzed in the field for dissolved oxygen (DO), electrical conductivity, temperature, and pH using field portable equipment. Data quality objectives and procedures for quality assurance and quality control were followed according to standard protocols for certified laboratories as specified in the California State University, Chico Environmental Laboratory's *Quality Assurance Project Plan.*

RESULTS AND DISCUSSION

Preliminary results for the four RTF retrofit systems (and the Oregon RTF system) are shown in Table 1. Performance of the RTF is evaluated by comparison of mean total N concentrations in septic tank effluent prior to RTF installation (w/o RTF) and after installation and acclimation of the RTF (w/RTF). As shown in Table 1, two of the RTF retrofits (Sites 4 and 5) are effectively removing a high percentage of the wastewater total N. Sites 4 and 5 are characterized by relatively high recirculation ratios (46 and 41, respectively), and both systems have documented consistently low effluent total N after a 3 to 4-week acclimation period (Figure 2). Dissolved oxygen levels with the RTF system averaged 3.4 mg/L at Site 4 and 1.2 mg/L at Site 5, but there is only a limited correlation of DO levels and NO_3^--N levels. The relatively high DO levels at Site 4 did not impede denitrification. Site 1 is believed to be constrained by the shorter trickling filter tower, which may have limited the development of a nitrifier population (U.S. EPA 1993). Site 3, serving restrooms in an office building, consistently nitrified 80 to 95% of the N entering the system, but denitrification apparently was limited by the low C:N ratio of the wastewater. Based on the stoichiometry of the generalized denitrification reaction, 2.86 mg

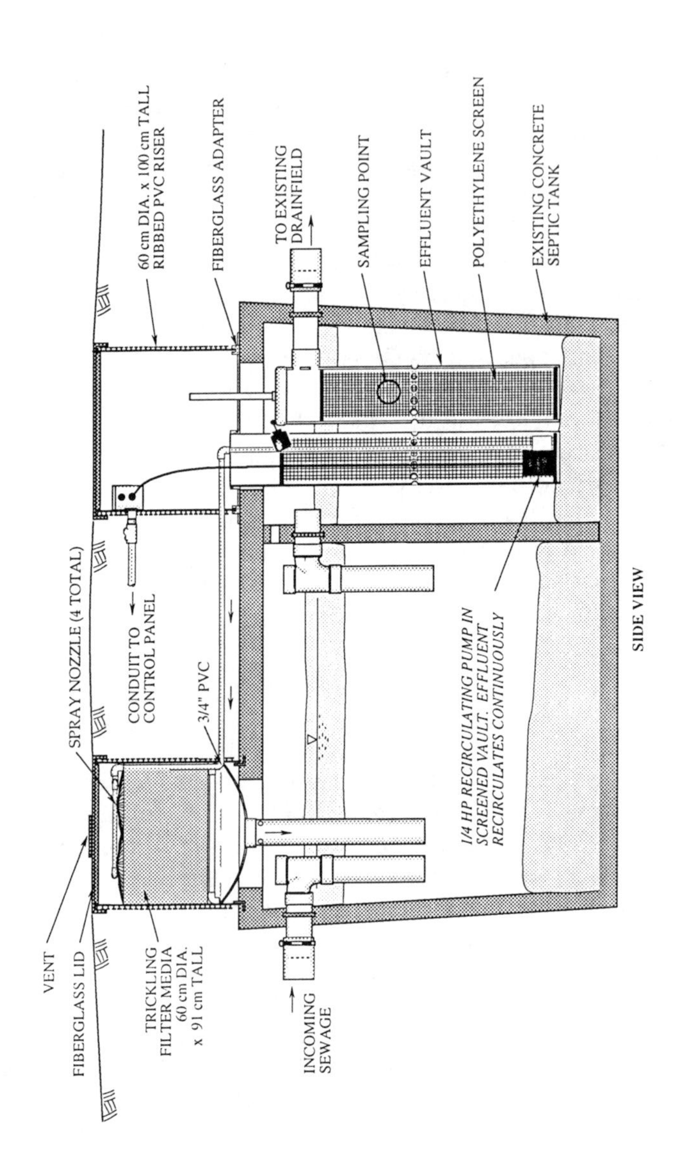

FIGURE 1. Recirculating trickling filter retrofit to existing two-compartment concrete septic tank (courtesy Orenco Systems Inc.).

TABLE 1. RTF retrofit system characteristics and performance data.

Site No.	Type (Occupancy)	Liquid Volume (L)	Filter Media Height (cm)	Q_{ave} (L/day)	R_T (–)	Mean Total N w/o RTF (mg/L)	Mean Total N w/RTF (mg/L)	Mean RTF N-Removal Efficiency (%)
1	SF (2 Adults and 2 Children)	3,700	61	1,000	31	28	24 ± 4[a]	20%
3	C (20 Adults)	7,570	91	300	90	24	50 ± 9	n/a
4	SF (2 Adults)	5,680	91	400	46	37	8 ± 2	80%
5	SF (2 Adults and 5 Children)	4,730	91	780	41	70	10 ± 5	86%
Ball (1994)	SF (2 Adults)	3,780	91	440	30	68	16	76%

SF : Single-family residence; C : Commercial; Q_{ave} : average daily wastewater flow; R_T : recirculation rate; a: mean value ± standard deviation

of oxygen demand is required to denitrify each milligram of NO_3-N (U.S. EPA 1993). A substantial level of organics removal was also documented for all four RTF systems with effluent BOD ranging from 10 to 49 mg/L compared to typical septic tank effluent levels of 129 to 147 mg/L (Canter and Knox 1985).

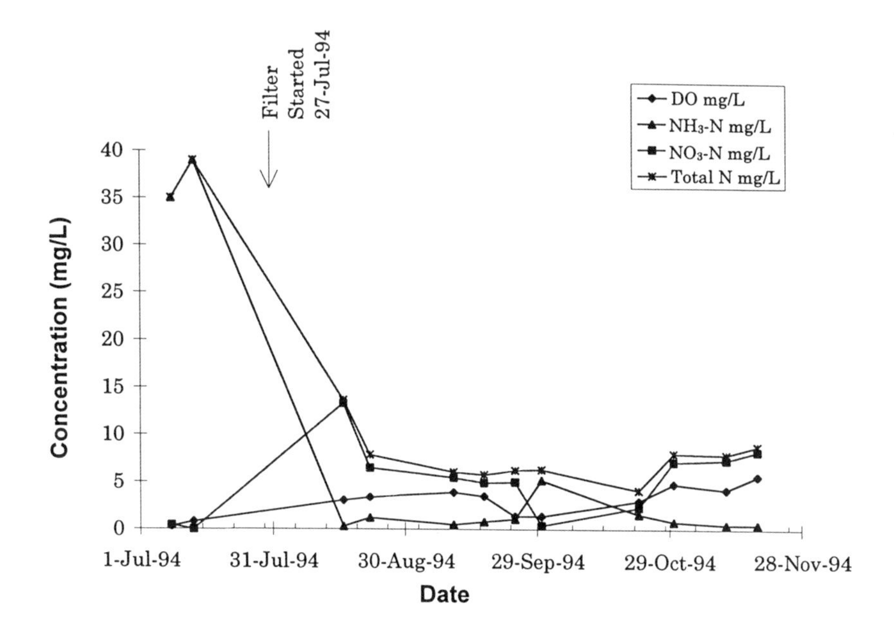

FIGURE 2. Nitrogen speciation and dissolved oxygen in RTF/septic tank effluent – Site 4.

TABLE 2. Comparison of RTF system to other on-site biological nitrogen removal systems.

Study	System	Numbers of Systems	Mean Total N in Effluent (mg/L)	Mean Total N Removal Efficiency (%)	Reference
Butte County, CA	RTF	3	12 ± 7	66 ± 29[a]	this work
Roseburg, OR	RTF	1	16	76	Ball 1994
University of Rhode Island	RUCK	3	17	56	Lamb et al. 1991
New Jersey	RUCK	15	20	54	Windisch 1990
Placer County, CA	ISF[b]	30	37	40	Cagle and Johnson 1994
Paradise, CA	RFG[c]	2	30	49	Nolte 1992

(a) Mean value ± standard deviation.
(b) ISF: intermittent sand filter.
(c) RGF: recirculating gravel filter.

The levels of nitrogen removal achieved by the RTF systems at Sites 1, 4, and 5 compare favorably with other on-site biological nitrogen removal (BNR) systems as shown in Table 2. Perhaps the most noteworthy feature of the RTF system is that primary, secondary, and tertiary wastewater treatment are all provided by a single unit. The cost of materials and labor for an RTF retrofit is estimated to range from $1,500 to $2,000; considerably lower than most sewer system connection fees.

CONCLUSIONS

Preliminary data on the performance of RTF systems retrofit to existing septic tanks document the potential for significant total nitrogen and BOD removal. Definition of optimum recirculation rates to maximize nitrogen removal may be a function of system BOD loading and is under study.

ACKNOWLEDGMENTS

This work was funded by the citizens of County Service Area 114, Butte County, California and the California State Water Resources Control Board. The authors wish to thank Mr. Ron Dykstra of the California Regional Water Quality Control Board and Dr. George Tchobanoglous of the University of California, Davis for providing peer review of the manuscript.

REFERENCES

Ball, H. L. 1994. "Nitrogen Reduction in an On-Site Trickling Filter/Upflow Filter Wastewater Treatment System." Presented at 7th National Symposium on Individual and Small Community Sewage Systems, American Society of Agricultural Engineers. Atlanta, GA. December 11-13, 1994.

Cagle, W., and L. A. Johnson. 1994. *Monitoring Report on the Use of Intermittent Sand Filters in Placer County*. Prepared for the Placer County Department of Health and Medical Services, CA.

Canter, L. W., and R. C. Knox. 1985. *Septic Tank Effects on Ground Water Quality*. Lewis Publishers, Chelsea, MI.

Greenberg, A. E., R. R. Trussell, and L. S. Clesceri. 1985. *Standard Methods for the Examination of Waste and Wastewater*, 16th ed. American Public Health Association, Washington DC.

Laak, R., M. A. Parese, and R. Costello. 1981. "Denitrification of Blackwater with Greywater." *J. Environmental Engineering Division ASCE 107*:581-590.

Lamb, B. E., A. J. Gold, G. W. Loomis, and G. C. McKiel. 1990. "Nitrogen Removal for On-Site Sewage Disposal: A Recirculating Sand Filter/Rock Tank Design." *Transactions of the ASAE 33*(2):525-531.

Lamb, B. E., A. J. Gold, G. W. Loomis, and G. C. McKiel. 1991. "Nitrogen Removal for On-Site Sewage Disposal: Field Evaluation of Buried Sand Filter/Greywater Systems." *Transactions of the ASAE 34*(3):883-889.

Metcalf & Eddy. 1991. *Wastewater Engineering: Treatment, Disposal, and Reuse*. 3rd ed. McGraw-Hill, New York, NY.

Metcalf & Eddy. 1995. *Los Osos Wastewater Study: Task F - Report on Sanitary Survey and Nitrate Source Study*, Chico, CA.

Nolte & Associates. 1992. "Literature Review of Recirculating and Intermittent Sand Filter Operation and Performance." Unpublished data from the Town of Paradise, CA.

Oakley, S. M., and L. Zeis. 1994. *The Quality Assurance Project Plan*. California State University, Chico, CA.

San Luis Obispo County. 1994. *Los Osos/Baywood Park Nitrogen Study*.

U.S. EPA. 1993. *Manual: Nitrogen Control*. EPA/625/R-93/010.

Wilhelm, S. R., S. L. Schiff, and J. A. Cherry. 1994. "Biogeochemical Evolution of Domestic Waste Water in Septic Systems: 1. Conceptual Model." *Ground Water 32*(6):905-916.

Windisch, M. A. 1990. *An Assessment of the Nitrogen Removal Efficiency and Performance of RUCK Septic Systems in the New Jersey Pinelands, New Jersey Pinelands Commission*, New Lisbon, NJ.

Biotreatment of Acid Rock Drainage at a Gold-Mining Operation

Thomas Wildeman, James Gusek, John Cevaal, Kent Whiting, and Joseph Scheuering

ABSTRACT

Acid rock drainage (ARD) from a closed gold-mining operation in northern California was treated on the pilot-scale level to determine the technical feasibility of passive treatment. The drainage has a pH of 3.8, and concentrations of Cu, Fe, Mn, Ni, and Zn of 140, 290, 28, 0.93, and 40 mg/L respectively. Based on laboratory results, a pilot system was constructed that consisted of a steel container filled with a substrate volume that measured 2 × 3 × 12 m. The substrate mixture was equal amounts by weight of cow manure, fine-grained soil, and limestone gravel. Loading of the system was based on the estimate that 0.3 moles of sulfide per m^3 of substrate per day would be generated, and the inflow of heavy metals should not exceed the sulfide generated. Using these principles, the flow was set at approximately 800 mL/min. Over the course of 10 months, the pilot system achieved removal of Cu and Ni below the effluent standards of 1.0 and 0.7 mg/L. Zn concentrations averaged approximately 0.1 mg/L, compared with an effluent standard of 0.02 mg/L. For a majority of the operation time, the cell was overloaded because the sulfide production rate of the substrate was lower than expected.

INTRODUCTION

In 1990, two breakthroughs on wetlands treatment of acid rock drainage (ARD) significantly advanced research at The Colorado School of Mines: (1) the determination that the primary anaerobic removal process is sulfate reduction and sulfide precipitation mediated by microbes (Wildeman et al. 1994a); and (2) the loading of a sulfate-reducing system for ARD can be determined by the rule of thumb that the level of microbial sulfate reduction should always exceed the amount of sulfide precipitation (Machemer et al. 1993, Wildeman et al. 1993). These two findings allowed development of anaerobic wetland systems by a staged design process similar to other biotreatment schemes. In particular, because microbial processes dominated, laboratory studies to determine suitable substrates and sulfate-reducing inocula can be conducted with a reasonable degree of certainty. Then, because effective treatment relies upon a sulfate reduction slightly exceeding metal sulfide precipitation, bench- and pilot-scale experiments can determine the proper loading factors and how a treatment system can be adapted to on-site conditions.

This paper reports on the first pilot-scale passive treatment system that was constructed using these concepts. The site was the Grey Eagle gold mine in northern California, where it is necessary to treat a continuous flow of seepage from a tailings impoundment. The seepage also contains ARD from historic mining. The composition of the commingled wastewater stream is given in Table 1. The mine site is owned by Noranda Grey Eagle Mines Inc., which funded the development project.

TREATMENT DECISIONS

The results of the laboratory study, reported by Wildeman and others (1994b), were encouraging enough that it was decided to forego bench-scale experiments and immediately design and construct a small pilot cell. The most promising substrate mixture for sulfate reduction consisted of ⅓ manure, ⅓ planter soil, and ⅓ limestone secured from Colorado sources. The original substrate materials from northern California were acidic soils and sawdust and these were not suitable for an anaerobic treatment system. Consequently, it was necessary to make another search for local materials that gave good sulfate reduction results in laboratory studies. In addition, the inocula of sulfate-reducing bacteria (SRB) came from Colorado sources, and sources from northern California had to be located so that transportation costs would be minimized. Consequently, another

TABLE 1. Comparison of whole water samples in mg/L. The substrate/water combinations are from the laboratory study. The passive system is effluent from the pilot cell taken during the first month of operation.

	Original Water	Passive System	30 g Substrate 90 g Water	60 g Substrate 60 g water	Effluent Limits
Cd	0.088	<0.0005	<0.005	<0.005	0.01
Cu	140	0.42	0.025	0.13	1.0
Ni	1.0	0.049	0.05	0.04	0.7
Fe	290	14	0.31	2.5	0.3
Mn	28	17	0.63	0.99	None
Zn	40	0.07	0.021	0.17	0.02
$SO_4^=$	1,500	2,200	170	270	None

round of laboratory tests was performed to find suitable organic materials and sulfate-reducing inocula. Because the laboratory results showed that a base mix of manure, limestone, and planter's soil mix was effective in promoting the activity of SRB and removing contaminants, the search for a local substrate centered upon these materials.

The only readily available manure came from a dairy farm and appeared to be processed such that it contained primarily hay with very little soil or excrement. It tested positive as an inoculum for sulfate-reducing bacteria; however, the growth of bacteria was slow. In the final designation of the substrate for the pilot reactor, it was decided to use the processed manure for the substrate formulation, but to include 10% of raw manure from the barnyard as the bacterial inoculum.

PILOT PLANT STUDY

The passive treatment system module (PTSM), described by Cevaal and others (1993), consisted of influent piping, substrate, and effluent piping installed in a welded steel tank. The rectangular tank was 10.9 m long, 2.7 m wide, and 2.7 m high and open at the top. The interior of the tank was covered with a bituminous polymer coating to protect the steel from the corrosive ARD. The substrate consisted of equal masses of processed manure, 1-cm-diameter limestone, and dark, fine-grained soil. The three components were blended and added to the PTSM in such a way that the substrate was not allowed to fall more than 1 m to minimize compaction. For inoculum, a portion of the substrate was blended with the raw manure and placed in the PTSM at the ⅓ and ⅔ level. The ARD to be treated passed downward through the 2-m-thick substrate and was collected in the effluent piping located at the base of the tank. Flow through the system was controlled hydraulically by the elevation of the effluent pipe relative to the level of water within the module.

Operation

The loading of the PTSM was based on the limiting reactant concept for loading of a sulfate-reducing reactor (Machemer et al. 1993, Wildeman et al. 1993). The premise is that the rate of generation of sulfide by the bacteria must slightly exceed the rate of flow of heavy metals (Mn + Fe + Cu + Zn) into the reactor. Based on previous studies (Reynolds et al. 1991, Wildeman et al. 1993), the rate of sulfide generation was estimated to be 0.3 mol $S^{=}/m^3$/day. The metals concentration in the influent totaled ~800 mg/L. Therefore, using an average molecular weight of 55 g/mol for the metals in the drainage, this results in a metals influent concentration of 0.014 mol/L. Given the volume of the PTSM as 55 m^3, the total sulfide produced per day was estimated to be 16.5 mole. Therefore, the flow was set so that at 800 mL/min, the metals input would not exceed 16.5 mol/d.

To start the test, impoundment seepage was added until the substrate was thoroughly soaked and water breached the surface. Then, the system was left to incubate for 2 weeks. The sulfate in the seepage and the easily extracted organic compounds from the manure served as excellent nutrients for the SRB. Previous studies found that one week of incubation produced vigorous sulfate reduction (Bolis et al. 1992, Wildeman et al. 1993). In this case it was assumed that after 2 weeks the system was an active bioreactor. The initial flow was set at 800 mL/min, and the PTSM was continuously operated for 22 months from October 1991 through July 1993. During this time, the physical operation of the PTSM was free of problems.

For the first 10-month period, the analytical results on total constituents in the effluent are given in Figure 1. For Fe, there were often large deviations between total and dissolved concentrations. When this happened, the suspended solids in the effluent water were not black, which is indicative of sulfides, but orange, which is indicative of ferric hydroxides. For the other constituents, except for excursions in Cu at the beginning of the study and Zn at the end of the period, the concentrations of dissolved constituents were only slightly less than the concentrations of total constituents. During this time, the composition of the influent water was essentially the same as that given in Table 1, except in March when annual recharge of the aquifer increased concentrations of contaminants in the ARD by 10% to 70%. This increase can be seen for the total of heavy metals in Table 2. In Table 2, the analytical results of sulfate concentrations in the influent and effluent are presented. The molarity of heavy metals in the influent and

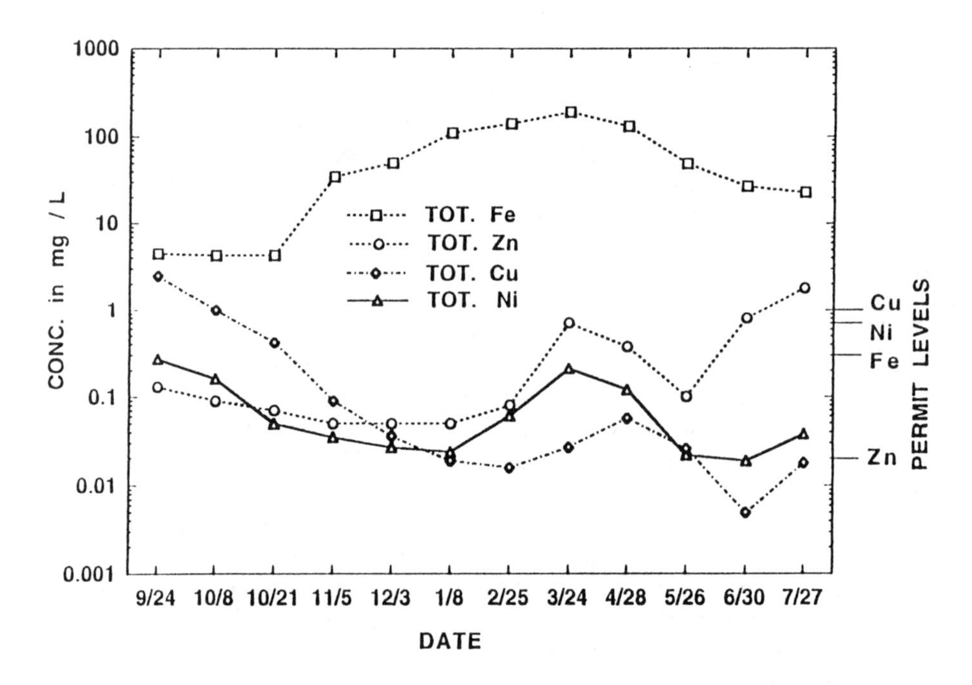

FIGURE 1. Unfiltered effluent metal concentrations in mg/L.

effluent are also shown so that the balance between sulfide produced and metals removed can be compared.

At the mine site, Eh, pH, and temperature measurements were performed twice a week as part of the on-site maintenance program. The PTSM tank was built of steel, was aboveground, and was not insulated. As a consequence, it responded rapidly to changes in ambient air temperature as the seasons changed. Figure 2 gives the on-site values of Eh and temperature for the effluent over the course of the 10-month study. Twice during the study, the temperature of the substrate at various depths and locations was determined. In both instances, the temperature range in the PTSM was less than 3°C with respect to both depth and lateral position.

Analysis of Operation

The metal concentrations of two laboratory samples and the PTSM effluent from 9/24/91 are compared in Table 1. For the first 6 weeks of operation, removal results were excellent and followed the results achieved in the laboratory study. However, beginning in November, the PTSM began to be overloaded. The results of this overloading can be seen in the rise in Fe concentration in Figure 1, and the rise in Eh in Figure 2. In Table 2, the imbalance in concentrations between sulfate reduced and heavy metals in the ARD also confirm that the cell was overloaded.

Cevaal and others (1993) and Wildeman and others (1994b) discuss this overloading and find that good immediate indicators of overloading are the rise in Eh above zero, and a change in the effluent is seen that may include suspended material such as ferric hydroxide. Good confirmation of overloading will be seen in a rise in the concentration of iron in the effluent. Also, the change in the reduction in sulfate concentration will be diminished and may even drop to zero (Wildeman et al. 1993).

The cause of the overloading is an inability to generate sulfide at the rate of 0.3 mol $S^{=}/m^3/day$. This inability is due to two possible causes. First, SRB activity is known to decrease with temperature decreases, and Kuyucak et al. (1991) found the most drastic decrease when the temperature fell below 10°C. In the PTSM, the temperature fell below 10°C on about November 15 and did not rise above 10°C until about April 15. Fe concentrations would be most affected by a reduction in SRB activity because FeS is the form of iron sulfide precipitated and it is the most soluble of the acid-volatile sulfides (Machemer et al. 1993, Wildeman et al. 1993).

Another cause for the inability to generate sulfide may be the formulation of the substrate. This was one of the first projects where laboratory results were used to design a substrate made of local materials. The laboratory studies were simple static tests and did not give an indication of how rapidly sulfate is being reduced. Subsequent studies would suggest boosting the organic content of the substrate. Bolis et al. (1992) performed bench-scale studies on two substrate mixes that were chosen from a series of laboratory studies to be the most promising.

TABLE 2. Sulfide generation in mol/L for the passive treatment system module based on sulfate reduction. For comparison, the metals removal in mol/L is also shown.

Date	Sulfate (mg/L)		Sulfide (Molar)	Metals (Mn+Fe+Ni+Cu+Zn+Cd) (Molar)		
	In	Out	Produced	In	Out	Removed
9/24/91	2,900	1,100	0.0188	0.0083	0.0002	0.0081
10/8/91	2,600	2,600	0.0000	0.0091	0.0001	0.0090
10/21/91	3,000	2,200	0.0083	0.0083	0.0005	0.0079
11/5/91	2,700	2,300	0.0042	0.0087	0.0006	0.0081
12/3/91	2,600	2,200	0.0042	0.0087	0.0008	0.0070
1/8/92	2,600	2,100	0.0052	0.0084	0.0015	0.0069
2/25/92	2,700	2,100	0.0063	0.0090	0.0019	0.0071
3/24/92	3,200	2,600	0.0063	0.0137	0.0027	0.0109
4/28/92	3,000	2,300	0.0073	0.0109	0.0015	0.0093
5/26/92	3,000	2,200	0.0083	0.0106	0.0005	0.0101
6/30/92	2,900	2,400	0.0052	0.0102	0.0004	0.0098

They found that a substrate of at least 75% composted manure with 25% soil or limestone performed at a reasonable level of sulfate reduction. In addition, they found that a "synthetic" substrate of 76% limestone, 14% alfalfa, and 10% inoculum by weight provided good sulfate reduction. Because of the difficulty in finding suitable substrate materials in this region, this "synthetic" mix would possibly be a better choice. In the substrate chosen for the Grey Eagle pilot cell, the dark, fine-grained soil is somewhat neutral with respect to buffering the ARD or providing organic nutrients for the SRB. Consequently, removing it and adding alfalfa and more limestone may provide a substrate capable of generating 0.3 mol sulfide/m^3/day on a long-term basis.

CONCLUSIONS

From this study, a number of conclusions can be made concerning the development of anaerobic wetlands and reactors to treat ARD. The limiting reactant concept of balancing sulfide production with metal inflow appears to control the operation of the PTSM. In addition, when not enough sulfide is generated, the concentration changes in the effluent can be predicted. Also, there appears to be some excess capacity in the reactor, so that if SRB activity is diminished or Eh increases, removal of Ni, Cu, and Zn will continue, but Fe removal may

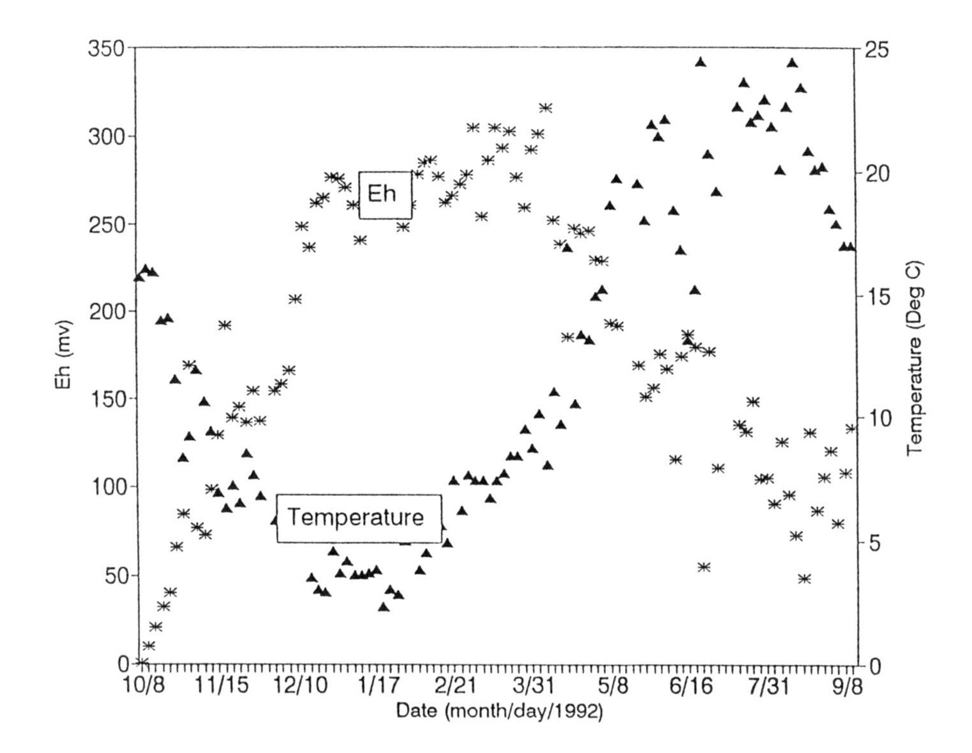

FIGURE 2. Variation of oxidation-reduction potential (Eh) with PTSM temperature in 1991 and 1992.

decrease. From the point of view of substrate development, various organic materials will work as organic nutrients for SRB. However, if the material chosen is significantly different from composted manure, then the sulfide production benchmark of 0.3 mol $S^{=}/m^3$/day may not be achieved. Finally, the expected relation between temperature, SRB activity, Eh, and contaminant removal has been established in the field.

ACKNOWLEDGMENTS

L. Destrampe of Noranda, David Updegraff of the Colorado School of Mines, and Roger Olsen and Susan Morea of Camp Dresser and McKee contributed significantly to the project.

REFERENCES

Bolis, J. L., Wildeman, T. R., and H. E. Dawson. 1992. "Hydraulic Conductivity of Substrates used for Passive Acid Mine Drainage Treatment." *Proceedings of the 1992 National Meeting of the American Society of Surface Mining and Reclamation.* Amer. Soc. Surf. Mining and Reclam., Princeton, WV. pp. 79-89.

Cevaal, J. N., R. L. Olsen, K. Whiting, T. R. Wildeman, and J. J. Gusek. 1993. "Passive biological treatment of acid drainage from a mining site: Laboratory and pilot scale evaluations." *Proceedings of the HMCRI Superfund XIV Conference.* Washington, DC.

Kuyucak, N., O. St.-Germain, and K. G. Wheeland. 1991. "In-situ bacterial treatment of AMD in open pits." *Proceedings of the Second International Conference on the Abatement of Acidic Drainage.* CANMET, Ottawa, Quebec. pp. 335-354.

Machemer, S. D., J. S. Reynolds, L. S. Laudon, and T. R. Wildeman. 1993. "Balance of S in a Constructed Wetland Built to Treat Acid Mine Drainage, Idaho Springs, Colorado, USA." *Appl. Geochem. 8*: 587-603.

Reynolds, J. S., S. D. Machemer, T. R. Wildeman, D. M. Updegraff, and R. R. Cohen. 1991. "Determination of the Rate of Sulfide Production in a Constructed Wetland Receiving Acid Mine Drainage." *Proceedings of the 1991 National Meeting of the American Society of Surface Mining and Reclamation.* Amer. Soc. Surf. Mining. and Reclam., Princeton, WV. pp. 175-182.

Wildeman, T. R., G. A. Brodie, and J. J. Gusek. 1993. *Wetland Design for Mining Operations.* BiTech Publishing Co., Vancouver, BC, Canada.

Wildeman, T. R., D. M. Updegraff, J. S. Reynolds, and J. L. Bolis. 1994a. "Passive Bioremediation of Metals from Water using Reactors or Constructed Wetlands." In J. L. Means and R. E. Hinchee (Eds.), *Emerging Technology for Bioremediation of Metals.* Lewis Publishers, Boca Raton, FL. pp. 13-24.

Wildeman, T. R., J. Cevaal, K. Whiting, J. J. Gusek, and J. Scheuering. 1994b. "Laboratory and Pilot-Scale Studies on the Treatment of Acid Rock Drainage at a Closed Gold-Mining Operation in California." *Proceedings of the International Land Reclamation and Mine Drainage Conference,* Vol. 2. U.S. Bureau of Mines Special Publication SP 06B-94. pp. 379-386.

Considerations in Deciding to Treat Contaminated Soils In Situ

Lawrence A. Smith and Janet M. Houthoofd

ABSTRACT

The development and application of in situ technologies for contaminated soil remediation is rapidly evolving. Diverse technologies are available for treatment of organics, inorganics, or a combination. Biological, chemical, and physical mechanisms can be applied to destroy, immobilize, or remove contaminants. These mechanisms can be either applied to excavated soil or used in situ. The costs, logistical concerns, and regulatory requirements associated with excavation, ex situ treatment, and disposal can make in situ treatment an attractive alternative. Treatment of contaminants in situ has both unique advantages and challenges. The obvious advantage is that no bulk excavation is required for in situ treatment. It must be recognized, however, that conditions in the subsurface will never be as controlled as in an ex situ process. Efficient delivery and recovery of treatment agents is essential to the effective application of in situ treatment. This paper describes factors to consider when selecting in situ treatment and outlines some characteristics of specific technologies. Both established and innovative technologies for remediation of contaminated soil are considered in the discussion, with emphasis on general factors influencing the effectiveness, implementability, and cost of in situ treatment.

INTRODUCTION

This paper provides information to assist personnel at contaminated sites in comparing in situ bioremediation, bioventing, and other in situ technologies. In situ technologies treat the hazardous constituents of a waste or contaminated environmental medium in place, without excavation. The technology must be capable of reducing the risk posed by these contaminants to an acceptable level (U.S. EPA 1990a).

Basic biological, chemical, and physical mechanisms can treat contaminants in both excavated and in situ soils. The costs and logistical concerns associated with excavation, ex situ treatment, and disposal can make in situ treatment an attractive alternative. In situ treatment entails the use of chemical or biological agents or physical manipulations to degrade, remove, or immobilize contaminants without requiring bulk soil removal. Typical in situ technologies include solidification/stabilization (S/S), soil vapor extraction (SVE), bioremediation/bioventing, vitrification, radiofrequency heating, soil flushing, and steam or hot air injection/extraction (U.S. EPA 1993a). Final technology selection should be based on site-specific evaluation and treatability testing (U.S. EPA 1992a). Assessing the feasibility of in situ treatment and selecting appropriate in situ technologies requires an understanding of the characteristics of the contaminants, the site, and the technologies, and of how these factors and conditions interact to allow effective delivery, control, and recovery of treatment agents and/or the contaminants.

DELIVERY AND RECOVERY OF TREATMENT AGENTS

Applying the treatment agents to the in situ geology gives in situ treatment unique advantages and challenges. The obvious advantage is that no bulk excavation is required for in situ treatment. Preventing excavation eliminates the cost and environmental consequences of moving the contaminated material. The conditions of the subsurface will never be as controlled as in an ex situ reactor, however. As a result, in situ treatment requires more extensive site characterization both before and after treatment, is harder to simulate in the laboratory, and must be designed and operated to minimize the spread of contamination.

The principal feature of in situ treatment is controlled delivery and recovery of energy, fluids, or other treatment agents to the subsurface. The treatment agent usually is water, air, or steam but may be energy input by conduction or radiation. For both physical- and energy-based in situ treatment agents, controlled application is a key to success. Systems must be available to apply treatment agents and to control the spread of contaminants and treatment agents beyond the treatment zone.

Efficient delivery and recovery methods control the effectiveness, implementability, and cost of in situ treatment. An array of delivery techniques are available to apply or inject treatment fluids into the subsurface. The types of delivery systems for in situ treatment can be classed generally as gravity driven, pressure driven, auger mixing, and energy coupling. Recovery systems typically fall in the gravity-driven or pressure-driven classification (U.S. EPA 1990b).

With the exception of radiofrequency heating and in situ vitrification, the in situ treatment methods cited above require delivery and control of liquid, slurry, gas, vapor, or a combination in the soil. For some technologies, the fluids also must be recovered after passing through the contaminated in situ volume. Fluid delivery may be accomplished by conventional gravity infiltration through surface or trench application or by pressure injection through wells. Conventional recovery methods include trenches or wells. The conventional methods rely on flow patterns determined by the design and placement of the drains or wells and the subsurface stratigraphy. As described below, innovative techniques are available to modify the subsurface conditions to improve flow rates or flow control.

In situ treatment agents include fluids delivered to the contaminated volume. Possible treatment fluids include hot gasses or vapors; water; or water-containing nutrients, surfactants, anions, cations, bacteria, S/S binder, or other treatment agents. For a technology to be effective, implementable, and economically competitive, the treatment agents must be delivered in a well-controlled manner. Conventional gravity- and pressure-driven methods are available to deliver and recover fluids. Gravity-driven methods rely on infiltration and collection due to hydraulic gradients. Typically delivery is by surface distribution and collection is by trench or similar drains. Pressure-driven methods rely on pressure gradients supplied by a source pump, a blower or steam generator, or an extraction pump or blower. A system of wells typically is used for delivery and recovery. The conventional delivery and recovery systems are highly dependent on the physico-chemical environment in the subsurface.

Innovative approaches are being developed and tested to improve the performance of delivery and recovery technologies in low-conductivity or heterogeneous geologic settings. The innovative delivery and recovery technologies may be devised to increase the conductivity in the treatment zone, decrease the conductivity below the treatment zone, or improve the efficiency of contact between the treatment agents and the material to be treated. Conductivity modification technologies include hydraulic fracturing, pneumatic fracturing, radial well drilling, jet slurrying, and kerfing. Technologies to improve the distribution or application efficiency of treatment agents include colloidal gas bubble (aphron) generation, ultrasonic methods, and cyclic pumping or steaming (U.S. EPA 1990a).

Auger-mixing technologies have been developed to deliver treatment agents with less reliance on a favorable existing geology. Auger mixing is applicable to delivery but not to recovery of treatment agents. The main examples of auger delivery are steam injection and addition and mixing of solidification/stabilization binders with augers. One vendor is testing auger mixing for addition of bioremediation nutrients.

Technologies to apply energy rather than fluids also are available for in situ treatment. Energy delivery systems reduce dependence on in situ conductivity but are sensitive to other in situ parameters. The key to energy delivery is good coupling of the electric or electromagnetic field to the soil being heated. The electric properties change as the moisture content changes. The energy input processes vaporize water so the electrical coupling properties of the soil must change as treatment proceeds. The changing soil properties increase the challenge in designing an efficient energy application system.

Systems for pneumatic fracturing and hydraulic fracturing to improve subsurface conductivity and systems to inject oxygen microbubbles to remediate groundwater in situ have both been accepted in the Superfund Innovative Technology Evaluation (SITE) Program. The demonstration of a pneumatic fracturing system was completed at a site located in South Plainfield, New Jersey

(Mack and Aspan 1993) and site demonstration documents are available (U.S. EPA 1993b).

Several in situ technologies rely on the ability to recover the treatment agent and contained contaminants from the subsurface. For example, recovery of contaminants containing flushing fluid is an integral part of soil flushing, and the collection and treatment of steam and condensate are essential to steam/hot air injection and extraction treatment.

General Factors in Selection of In Situ Technology

Factors to consider in the selection of in situ treatment methods and during evaluation of in situ technologies include the general technology capabilities and generic factors that influence the suitability of in situ treatment compared to ex situ treatment. The Superfund process for screening and selecting technologies is described by the U.S. EPA (1988a). Preliminary screening of technologies is based on effectiveness, implementability, and cost. Although this paper describes general features to consider when selecting in situ treatment, the user should keep in mind that ex situ technology candidates should be considered along with the in situ options. At many sites, both in situ and ex situ technologies may be competing candidates late in the technology selection process.

Selecting a technology often requires several iterations with increasingly well-defined data to refine the selection. As the project progresses, technology-specific and site-specific information becomes available. This information is used to better define which technologies are suitable for waste materials and conditions at the site. As the decision maker obtains more information about site conditions, waste characteristics, and treatability study results, this paper can be used to help further refine selection of candidate technologies. However, as the list of candidates is narrowed, additional published sources and expert opinion should be sought to obtain more detailed information about the candidate technologies.

TECHNOLOGY CHARACTERISTICS

The applicability of the technology to the contaminants present, the technology maturity, and the ability of the technology to operate in the unsaturated and/or saturated zones should be considered in technology selection. The applicability of the technology to general types of contaminants is summarized in Table 1. The characteristics of the technologies are summarized in Table 2. Preliminary selection of technology candidates can be based on the capabilities of in situ technologies to treat chemical groups present at the site.

The chemical contaminant groups considered are divided into three general groups: organics, inorganics, and reactives. The types of organics considered are halogenated and nonhalogenated volatile organic compounds (VOCs), halogenated and nonhalogenated semivolatile organic compounds (SVOCs),

TABLE 1. Effectiveness of in situ treatment on general contaminant groups for soil.

	Contaminant Groups	In Situ Solidification/ Stabilization (a) (b)	Soil Vapor Extraction (b) (c)	In Situ Bioremediation (d)	Bioventing (e)	In Situ Vitrification (d) (f)	Radio-frequency Heating (f)	Soil Flushing (b) (g)	Steam Injection: Stationary System (b) (f) (h)	Steam Injection: Mobile System (b) (f) (h)
Organic	Halogenated volatiles	X[1]	■	◆	□	◆	◆	■	■	■
	Halogenated semivolatiles	◆[2]	◆	◆	◆[4]	◆	◆	◆	◆	◆
	Nonhalogenated volatiles	X[1]	■	◆	■	◆	■	◆	■	■
	Nonhalogenated semivolatiles	◆[2]	■	◆	■	◆	■	■	◆	◆
	PCBs	◆	□	◆	□	◆	◆	◆	◆	□
	Pesticides	◆	□	◆	□	◆	◆	◆	◆	□
	Dioxins/Furans	◆	□	◆	□	◆	□	◆	◆	□
	Organic cyanides	◆	□	◆	□	◆	□	◆	◆	□
	Organic corrosives	◆	□	X	X	◆	□	◆	◆	□
Inorganic	Volatile metals	■[3]	□	X[5]	X[5]	◆	□	◆	◆[6]	□
	Nonvolatile metals	■	□	X[5]	X[5]	◆	□	■	◆[6]	□
	Asbestos	■	□	□	□	◆	□	□	□	□
	Radioactive materials	■	□	X	X	■	□	◆	◆[6]	□
	Inorganic cyanides	■	□	X	X	◆	□	◆	◆[6]	□
	Inorganic corrosives	■	□	X	X	◆	□	◆	◆[6]	□
Reactive	Oxidizers	◆	□	X	X	◆	□	◆	◆[6]	□
	Reducers	◆	◆	X	X	◆	□	◆	◆[6]	□

■ Demonstrated effectiveness: Successful treatability test at some scale completed. □ No expected effectiveness: No mechanistic basis indicating that technology will work.
◆ Potential effectiveness: Mechanistic basis indicating that technology will work. X Potential adverse effects.

(1) Vaporization and emission of volatile organic compounds may pose a hazard during mixing. (2) Semivolatile organics are difficult to treat, but low concentrations of some compounds can be treated. (3) Arsenic and mercury are difficult to immobilize with cement-based binder formulations. (4) Possible to treat by cometabolism techniques. (5) Metals can interfere with bioremediation or bioventing of organics; however, bioremediation methods for low concentrations of metals are being developed. (6) Potential effectiveness only for water-soluble compounds.

Adapted from the following sources: (a) U.S. EPA 1993c, EPA/530/R-93/012. (b) Donehey et al. 1992. (c) U.S. EPA 1991b, EPA/540-2-91/006. (d) U.S. EPA 1988b, EPA/540/2-88/004. (e) U.S. Air Force 1992. (f) Houthoofd et al. 1991, EPA/600/9-91/002. (g) U.S. EPA 1991a, EPA/540/2-91/021. (h) U.S. EPA 1991c, EPA/540/2-91/005.

TABLE 2. Summary of in situ technology characteristics.

Technology/Maturity	Media Typically Treated	Typical Agents or Amendments	Delivery Methods	Recovery Methods
Solidification/stabilization (a)(b) Established	Saturated or unsaturated soil, sediment, or sludge Approximate depth limit: 30 ft (9 m) for auger, several feet for in-place mixing, and not a major constraint for grout injection	Cement, fly ash, blast furnace slag, lime, or bitumen	Auger mixing, in-place mixing, or injection	None required
Soil vapor extraction (b)(c) Innovative	Unsaturated soil	Air	Passive air inlet or injection wells	Air extraction wells (off-gas treatment may be required)
In situ bioremediation (d) Innovative	Saturated or unsaturated soil, sediment, or sludge	Aqueous solution containing an electron acceptor (typically oxygen), nutrients, pH modifiers, or additives	Surface infiltration, tilling, or water injection wells	None required
Bioventing (e) Innovative	Unsaturated soil, sediment, or sludge	Air	Passive air inlet or injection wells	Air extraction wells may be used (off-gas treatment may be required)
In situ vitrification (d)(f) Innovative	Saturated or unsaturated soil, sediment, or sludge Approximate depth limit 20 ft (6 m) with possible extension to 30 ft (9 m)	Electrical energy by conduction	Electrodes	Off-gas collection and treatment
Radiofrequency heating (f) Innovative	Unsaturated soil, sediment, or sludge	Electrical energy by radiation	Radiofrequency antennae system	Off-gas collection and treatment
Soil flushing (b)(g) Innovative	Unsaturated or saturated soil	Water, acidic solutions, basic solutions, chelating agents, or surfactants	Extraction fluid injection wells	Extraction fluid recovery wells
Steam/hot air injection stationary system (b)(f)(h) Innovative	Saturated or unsaturated soil	Steam and/or hot air	Steam injection wells	Condensate recovery wells and off-gas collection and treatment
Steam/hot air injection mobile (auger) system (b)(f)(h) Innovative	Saturated or unsaturated soil Approximate depth limit: 30 ft (9 m) for auger system	Steam and/or hot air	Auger mixing	Off-gas collection and treatment

Adapted from the following sources: (a) U.S. EPA 1993c. (b) Donehey et al. 1992. (c) U.S. EPA 1991b. (d) U.S. EPA 1988b. (e) U.S. Air Force 1992. (f) Houthoofd et al. 1991. (g) U.S. EPA 1991a. (h) U.S. EPA 1991c.

polychlorinated biphenyls (PCBs), pesticides, dioxins and furans, organic cyanides, and organic corrosives. Inorganics are subdivided into volatile metals (and metalloids), nonvolatile metals, asbestos, radioactive materials, inorganic cyanides, and inorganic corrosives. Reactive species may be either oxidizers or reducers. More detailed lists of constituents within each contaminant group are given by the U.S. EPA (1988b).

Treatment often requires a sequence of operations, or treatment train, to deal with a combination of wastes. When evaluating wastes containing contaminants from more than one chemical constituent group, each waste group initially should be considered separately to develop a list of potentially applicable treatment technologies for each chemical group present in the soil. The technology lists can be compared to determine if some candidate technologies are able to treat all of the groups present.

If one technology is unable to treat all of the groups, development of a treatment train may be required. For example at a site with a combination of VOCs and metal contaminants, soil vapor extraction (SVE) can be used to remove the VOCs followed by in situ solidification/stabilization to reduce the mobility of the metals. The selected treatment train also must be reviewed for potential interferences or adverse effects. For example, SVE may increase the proportion of hexavalent chromium, increasing the mobility and toxicity of the chromium.

One of the following four characteristics is indicated for each combination of contaminant group and in situ technology in Table 1:

1. **Demonstrated Effectiveness** — The technology has been shown to treat some contaminants in the chemical group to acceptable levels when applied to contaminated soil. Treatment may involve removal, destruction, immobilization, or toxicity reduction. The demonstration may have been at the laboratory, pilot, or production scale.
2. **Potential Effectiveness** — Literature reports indicate a mechanistic basis for expecting that the technology will remove, destroy, immobilize, or otherwise treat some of the chemicals in the group when used to treat soil.
3. **No Expected Effectiveness** — There is no known mechanistic basis for expecting that the technology will remove, destroy, immobilize, or otherwise treat some of the chemicals in the group when used to treat soil.
4. **Possible Adverse Effects** — The contaminant is likely to interfere with the treatment technology or to adversely affect safety, health, or the environment. Adverse effects may occur only when the contaminant concentration is above a threshold level. In many cases, the adverse effect may be alleviated by pretreatment to reduce the concentration of the adverse contaminant.

Table 2 indicates the maturity of the technology and its applicability for saturated and unsaturated media. The maturity is indicated by the ranking shown below (U.S. EPA 1992b). Technology maturity is an important factor in the cost and timeliness of technology implementation.

1. **Established Technology** — The technology has been used on a commercial scale and has been established for use in full-scale remediations (e.g., incineration, capping, solidification/stabilization).
2. **Innovative Technology** — The technology is an alternative treatment technology (i.e., "alternative" to land disposal) for which use at Superfund-type sites is inhibited by lack of data on cost and performance.

To further assist in the review of technology candidates, Table 2 indicates the media typically treated, typical treatment agents or amendments, and delivery and recovery methods. Figure 1 shows the approximate range of in situ remediation costs. The costs shown are based on limited data reported in the literature. The sources rarely give full characterization of elements included in the cost estimates. The ranges should be viewed as preliminary indications of approximate comparative costs of the various technologies.

GENERIC FACTORS FOR FEASIBILITY SCREENING OF IN SITU TREATMENT

Several factors apply to the evaluation of in situ treatment at most sites. These generic factors have broad application regardless of the specific technology. Five categories have been identified to assist in organizing consideration of the potential feasibility of the in situ treatment for a particular site. This evaluation relates to the three screening criteria named in the National Contingency Plan (NCP) instituted by the Comprehensive Environmental Response, Compensation, and Liability Act (CERCLA) of 1980 and described by the U.S. EPA (1988a): effectiveness, implementability, and cost. The five categories are described in Table 3.

These generic factors give an overall framework for evaluating the potential for using in situ technologies. Site conditions that give a poor ranking in one or even several factors do not necessarily indicate that in situ approaches are unlikely to succeed. The generic factors are geologic and waste material characteristics that are significant in controlling or affecting the effectiveness or implementability of in situ technologies. Although these factors generally are of interest at all sites, some have more effect on the performance of specific technologies. The user must not draw a conclusion that in situ treatment is inappropriate based on one or two unfavorable factors. The design features of a particular technology may be able to eliminate or avoid some of the limitations inherent with most in situ treatment technologies. For example, in situ solidification/stabilization (S/S) technologies using mechanical mixing are less affected by the initial soil conductivity than are technologies that require delivery of fluid flow (see Table 2). Moreover, technologies such as steam injection, in situ vitrification, and radiofrequency heating, although generally slower than conventional ex situ methods, can proceed more quickly than in situ bioremediation or soil vapor extraction.

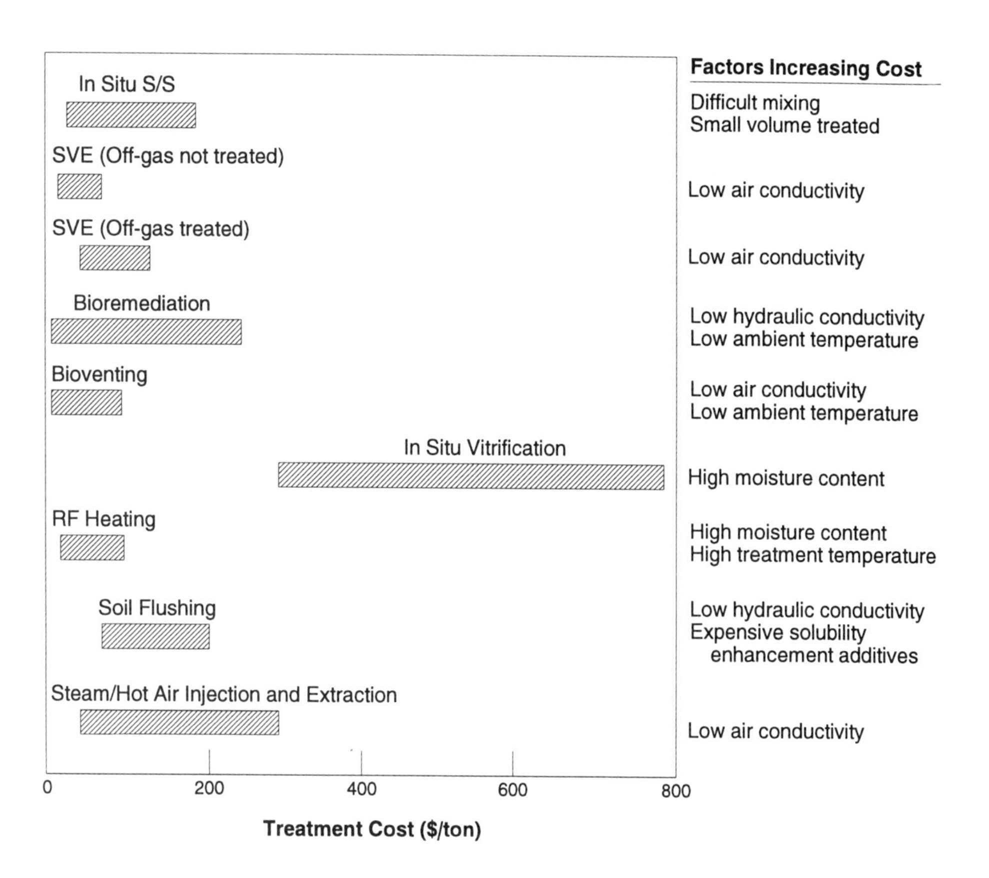

FIGURE 1. Estimated cost ranges of in situ remediation technologies.

For in-depth evaluation of technology alternatives, the consideration of these broad, generic factors must be supplemented by the consideration of many specific factors that are critical to the success of individual technologies. Factors important for individual technologies vary greatly. They would include items such as the moisture content or temperature of the soil or the VOC content in the contaminated soil.

Hydrogeologic Flow Regime

The hydrogeologic flow regime characterizes the gas and liquid flow in the subsurface. Examination of flow regime characteristics is directed at answering questions such as the following:

- Will contaminant removal be achieved at an acceptable rate?
- Will contaminant removal be complete and uniform?
- Will contaminants or treatment agents escape from the treatment area?

TABLE 3. Generic factors influencing selection of in situ treatment.

Factor Influencing Selection of In Situ Treatment	Conditions Favoring Selection of In Situ Treatment	Basis
Hydrogeologic flow regime	High or moderate conductivity uniformly distributed in formation or Low-conductivity regions surrounded by regions of high or moderate horizontal conductivity (a)	Treatment reagents must reach contaminated matrix by advective or diffusional flow.
	Deep water table and/or competent aquitard below contaminated volume	Delivery and recovery of treatment agents must be controlled.
Regulatory standards	Wastes that are difficult to treat in accordance with Land Disposal Restriction (LDR) requirements	LDRs apply to excavated material, unless the material is excavated and treated within a Corrective Action Management Unit (CAMU).
Time available for remediation	Completion time not critical	In situ treatment requires more time to complete than ex situ treatment.
Removal logistics	Large volume of waste Waste not accessible due to existing structures Excavation difficult due to matrix characteristics or depth Poor transportation infrastructure	In situ treatment does not require excavation.
Waste conditions	Large volume of waste	It is not economical to excavate large volumes for treatment of low concentrations.
	Low contaminant concentrations	In situ treatment may reduce the need for capital-intensive treatment equipment.

(a) The low-conductivity regions must be "thin" with respect to diffusion path length, which can be feet or inches for gas-phase diffusion in dry soils and inches or less for water-phase diffusion. (See Table 2 for information on type of treatment agent.)

The flow regime factor is controlled mainly by the amount of available primary and secondary fluid flow routes, the magnitude and homogeneity of hydraulic conductivity, fluid levels and pressures, and the proximity to a discharge location. Information needed to define the hydrogeologic flow regime includes a complete understanding of the geologic strata and how they were deposited, full characterization of the fluids and deposits for fluid transmission properties, and monitoring of soil moisture and water levels through at least three seasons of 1 year.

Geologic, hydraulic, and fluid-behavior data are needed to evaluate the flow regime. Geologic data include, in part, primary and secondary effective conductivity, history of geologic strata formation, and the stratigraphic and structural characteristics of the deposit. Hydrologic data include both surface water and groundwater flow, level, and pressure characteristics. Surface water data, such as stream/lake hydrographs and precipitation, infiltration, and recharge measurements, are needed to understand the general water balance of the system, whereas groundwater data, including pressure graphs, well hydrographs, and hydraulic conductivity and dispersion measurements, are needed to calculate water and mass flux through the system.

Understanding the spatial variation of conductivity also is essential to evaluate candidate in situ treatment technologies. Preferred flow pathways develop in the subsurface due either to inhomogeneities in the conductivity or to geologic facies. Most soils have preferential flowpaths that are responsible for much of the conductivity. The preferential paths can arise from a number of causes such as root intrusions, shrink/swell or wet/dry cycling, or uneven settling (U.S. EPA 1990c). These preferred pathways result in high hydraulic conductivity contrasts that can diminish the reliability and efficiency of in situ treatment methods. Geologic deposits where the variation of hydraulic conductivity with direction is less than a factor of 10 will be conducive to in situ methods. For most efficient application of in situ treatment, the deposits should have little or no vertical fracturing and no highly developed bedding planes. Implementation time will be less and removal will be more complete when the conductivity tends toward homogeneity.

Uniformly high conductivity in the contaminated media is best for application of in situ treatment. Hydraulic conductivity of more than 10^{-3} cm/s is most favorable to technologies that require flow of water solutions. For technologies that require air or vapor flow (see Table 2), an air conductivity of more than 10^{-4} cm/s is most favorable (U.S. EPA 1990c). In situ treatment still can be applied in geologies with much lower conductivities. However, contaminant transport in the lower conductivity regions will occur by slower diffusion processes rather than by bulk material flow. Feasibility depends on the type of treatment agent, the contaminant transport mechanisms, and the details of the distribution of the primary and secondary flowpaths.

Many in situ treatment technologies require injection of treatment agents such as steam, chemicals, or nutrients. Often the treatment agents must then be collected from the subsurface for further processing. The subsurface geology should be amenable to containment of the treatment agents in the contaminated area. Containment will be maximized when vertical and horizontal hydraulic gradients are low or if the treatment zone is bounded geologically by deposits with low hydraulic conductivity. Close proximity to groundwater discharge areas such as streams, lakes, and seeps can jeopardize containment of in situ treatment agents.

Regulatory Standards

The regulatory standards factor characterizes the overall regulatory climate at the site based on federal, state, and local regulations. Examination of regulatory standards is directed at answering questions such as these:

- What contaminant cleanup levels are required?
- Are land-use restrictions consistent with the candidate technologies?
- Will in situ treatment cause unacceptable alteration of soil conditions?
- Is injection of treatment chemicals consistent with Land Disposal Restrictions (LDRs) and other regulations, as required?

In situ treatment may require more extensive sampling than ex situ treatment to demonstrate that required treatment performance levels have been achieved. With in situ treatment, the variation in natural conditions and the distribution of the contaminant must be determined. This often requires extensive sampling to build a statistical basis for evaluating whether or not analytical results represent in situ conditions. In contrast, in a typical ex situ treatment system, waste material is excavated, prepared, and homogenized as part of the treatment operation. These homogenized batches can be represented with a smaller number of samples than corresponding in situ materials.

Technologies, such as solidification/stabilization and some applications of bioremediation that accomplish the treatment in situ, may reduce point source air emissions or other discharges. Many in situ treatment technologies, however, do have aboveground components. For example, materials are injected; groundwater is extracted, treated, and reinjected; or vapors are captured and treated. The aboveground portion still may be subject to appropriate environmental regulations. Technologies that require injection of fluids may need to follow Underground Injection Control regulations.

Time Available for Remediation

The available time factor characterizes the amount of time allowed to set up, operate, and remove the treatment technology. Determining the time available to complete remediation is directed at answering questions such as:

- Can the cleanup be completed in a time frame consistent with health, safety, and environmental protection?

- Can the cleanup be completed in a time frame consistent with end-use requirements?

The time available for remediation is controlled first by the need to protect human safety and health and the environment. Remediation must proceed quickly if a toxic contaminant is present, the contaminant concentration is high, or the contaminant is mobile and near a critical ecosystem. Time available may be controlled also by the value or intended end use for the site. It is undesirable to hold a high-value site out of productive use for a long period.

In situ remediation typically requires more treatment time than the analogous ex situ treatment technology. In situ bioremediation, for example, typically requires about 4 to 6 years (U.S. EPA and U.S. Air Force 1993). Excavation allows essentially immediate remediation of the site. However, the excavated material often must be shipped and stored before treatment. Rapid remediation is needed if the contaminant presents an imminent danger due to hazard level, mobility, or other factors. Rapid remediation of an imminent hazard generally favors an ex situ remediation approach.

The importance of the length of remediation time may be lessened if the time constraint is driven by economic or end-use requirements. Many in situ technologies can be applied concurrently with other site operations. For example, well and injection/extraction equipment for bioventing, soil vapor extraction, or fixed-system steam injection do not occupy the full surface area of a site. Depending on the technology and the site use, it may be possible to continue routine site operations during an in situ remediation. However, the need for rapid remediation still generally increases the favorability of ex situ treatment technologies.

Removal Logistics

The removal logistics factor characterizes the feasibility of excavating, handling, and transporting the contaminated soil. Examination of removal logistics is directed at answering questions such as:

- Is the material accessible for excavation?
- Can the contaminated soil or water be moved efficiently by conventional bulk material-handling equipment and techniques?
- Will on-site (and if needed off-site) infrastructure support transport of waste materials?

Removal logistics are determined by access to the contaminated site for excavation, the ability to handle excavated materials, space for placement of ex situ treatment equipment, and the road and rail system on and around the site.

Data needed to evaluate the removal logistics include a map of the site showing the general arrangement of structures and infrastructure and an approximate assessment of the subsurface conditions such as the location of contamination and the location of major geologic and hydrogeologic features such as surface water and aquifers.

Poor removal logistics favor in situ treatment. In situ treatment generally is favored by conditions such as contamination located under a building that is to remain after remediation; presence of buried piping or utility lines in the area; contamination located at great depth or under a rock formation; poor road or rail access; nearby businesses, schools, or heavy traffic areas; or site location in a remote area distant from treatment facilities or sources of backfill. Contamination located deeper than 5 feet or occupying a volume of more than 1,000 m^3 increases both the cost and the complexity of excavation (U.S. EPA 1990c). Specialized delivery and recovery systems may be necessary to overcome poor site logistics.

Waste Conditions

The waste conditions factor characterizes the chemical and physical form of the waste with respect to the ability to effectively treat or remove the contaminant. Examination of waste conditions is directed at answering questions such as:

- Are the concentration and distribution of contaminants consistent with effective in situ treatment?
- Does the waste distribution or condition allow effective delivery of treatment agents to the contaminant?

The waste conditions factor is controlled by the in situ conditions of the contaminant and matrix. The conditions requiring characterization include the concentration and distribution of the contaminant, the chemical form and speciation of the contaminant and matrix, and physical properties of the waste and matrix.

Data needed to characterize the waste conditions include a survey of the location, concentration measurements, and a description of the form of contaminant, matrix, and debris in the remediation site. Some soil sampling data may be available, but assessment of the waste condition at the preliminary evaluation stage typically will be based largely on historical records.

The understanding of waste conditions must be constantly reevaluated as additional data are obtained. In addition to estimating the areal extent and concentration of contamination, the assessment must address the possibility of the contaminant being contained in drums or tanks and the potential presence of noncontaminant debris that could make excavation difficult or obstruct the flow of in situ treatment agents.

In general, contaminants that are either highly concentrated or spread over a relatively small area are best treated by ex situ methods. In particular, contaminants contained in drums or underground tanks are difficult to treat with in situ methods. Dilute or widely distributed contaminants tend to favor in situ treatment. When the contaminant is present at low concentration, ex situ processing requires excavation, handling, and, processing of a high proportion of matrix materials relative to a small amount of contaminant.

CONCLUSIONS

In situ treatment can be the best remediation approach at a site for a variety of reasons. In situ treatment avoids bulk excavation of contaminated material and can minimize disruption of the site and reduce public and worker exposure to contaminated materials. On the other hand, in situ methods must be applied in the existing hydrogeologic conditions. Frequently the existing in situ conditions are heterogeneous and provide less control of the treatment process than ex situ treatment. Evaluation of the applicability of in situ treatment and selection of individual technologies to remediate the environmental problems at a specific site require consideration of several factors. This paper provides an approach for structuring comparisons of in situ bioremediation and bioventing with other in situ treatment methods.

ACKNOWLEDGMENTS AND DISCLAIMER

This article was developed with the assistance of the U.S. Environmental Protection Agency Engineering Forum. The authors wish to recognize the contributions of the Engineering Forum contacts, Robert Stamnes and Paul Leonard, and the assistance of Joan Colson of U.S. Environmental Protection Agency's Risk Reduction Engineering Laboratory. Although the research described in this article has been funded wholly by the U.S. Environmental Protection Agency through Contract Number 68-C0-0003 to Battelle, it has not been subjected to Agency review. Therefore, it does not necessarily reflect the views of the Agency. Mention of trade names or commercial products does not constitute endorsement or recommendation for use.

REFERENCES

Donehey, A. J., R. A. Hyde, R. B. Piper, M. W. Roy, and S. S. Walker. 1992. "In Situ Physical and Chemical Treatments." *Proceedings of the 1992 U.S. EPA/A&WMA International Symposium on In Situ Treatment of Contaminated Soil and Water.* Air & Waste Management Association. Pittsburgh, Pennsylvania. pp. 98-106.

Houthoofd, J. M., J. H. McCready, and M. H. Roulier. 1991. "Soil Heating Technologies for In Situ Treatment: A Review." *Proceedings of the 17th Annual Hazardous Waste Research Symposium.* EPA/600/9-91/002. Office of Research and Development, Risk Reduction Engineering Laboratory, Cincinnati, Ohio. pp. 190-203.

Mack, J. P., and H. N. Aspan. 1993. "Using Pneumatic Fracturing Extraction to Achieve Regulatory Compliance and Enhance VOC Removal from Low-Permeability Formations." *Remediation*, 3(3):309-326.

U.S. Air Force. 1992. *Test Plan and Technical Protocol for a Treatability Test for Bioventing, Rev. 2.* U.S. Air Force Center for Environmental Excellence, Brooks Air Force Base, TX.

U.S. Environmental Protection Agency. 1988a. *Guidance for Conducting Remedial Investigations and Feasibility Studies under CERCLA — Interim Final.* EPA/540/G-89/004. Office of Emergency and Remedial Response, Washington, DC.

U.S. Environmental Protection Agency. 1988b. *Technology Screening Guide for Treatment of CERCLA Soils and Sludges.* EPA/540/2-88/004. Office of Emergency and Remedial Response, Washington, DC.

U.S. Environmental Protection Agency. 1990a. *Handbook on In Situ Treatment of Hazardous Waste-Contaminated Soils.* EPA/540/2-90/002. Office of Research and Development, Risk Reduction Engineering Laboratory, Cincinnati, Ohio.

U.S. Environmental Protection Agency. 1990b. *Technologies of Delivery or Recovery for the Remediation of Hazardous Waste Sites.* EPA/600/2-89/066; NTIS PB90-156 225/AS. Office of Research and Development, Risk Reduction Engineering Laboratory, Cincinnati, Ohio.

U.S. Environmental Protection Agency. 1990c. *Assessing UST Corrective Action Technologies: Site Assessment and Selection of Unsaturated Zone Treatment Technologies.* EPA/600/2-90/011. Office of Research and Development, Risk Reduction Engineering Laboratory, Cincinnati, Ohio.

U.S. Environmental Protection Agency. 1991a. *Engineering Bulletin — In Situ Soil Flushing.* EPA/540/2-91/021. Office of Emergency and Remedial Response, Washington, DC; and Office of Research and Development, Risk Reduction Engineering Laboratory, Cincinnati, Ohio.

U.S. Environmental Protection Agency. 1991b. *Engineering Bulletin — In Situ Soil Vapor Extraction Treatment.* EPA/540/2-91/006. Office of Emergency and Remedial Response, Washington, DC; and Office of Research and Development, Risk Reduction Engineering Laboratory, Cincinnati, Ohio.

U.S. Environmental Protection Agency. 1991c. *Engineering Bulletin — In Situ Steam Extraction Treatment.* EPA/540/2-91/005. Office of Emergency and Remedial Response, Washington, DC; and Office of Research and Development, Risk Reduction Engineering Laboratory, Cincinnati, Ohio.

U.S. Environmental Protection Agency. 1992a. *Guide for Conducting Treatability Studies under CERCLA, Final.* EPA/540/R-092/071a. Office of Research and Development, Risk Reduction Engineering Laboratory; and Office of Emergency and Remedial Response, Office of Solid Waste and Emergency Response, Washington, DC.

U.S. Environmental Protection Agency. 1992b. *Innovative Treatment Technologies: Semi-Annual Status Report*, 4th ed. EPA/542/R-92/011. Office of Solid Waste and Emergency Response, Technology Innovation Office, Washington, DC.

U.S. Environmental Protection Agency. 1993a. *VISITT Vendor Information System For Innovative Treatment Technologies. VISITT User Manual, Version 2.* EPA/542/R-93/001. Office of Solid Waste and Emergency Response, Washington, DC.

U.S. Environmental Protection Agency. 1993b. *The Superfund Innovative Technology Evaluation Program — Technology Profiles*, 6th ed. EPA/540/R-93/526. Office of Research and Development, Cincinnati, Ohio.

U.S. Environmental Protection Agency. 1993c. *Technical Resources Document on Solidification/Stabilization and Its Application to Waste Materials.* EPA/530/R-93/012. Risk Reduction Engineering Laboratory, Cincinnati, Ohio.

U.S. Environmental Protection Agency and U.S. Air Force. 1993. *Remediation Technologies Matrix Reference Guide.* Draft report. U.S. Environmental Protection Agency, Technology Innovation Office, Washington, DC; and U.S. Air Force, Armstrong Laboratory Environics Directorate, Tyndall Air Force Base, Florida.

AUTHOR LIST

Aronstein, Boris N.
Institute of Gas Technology
1700 South Mt. Prospect Rd.
Des Plaines, IL 60018 USA

Bachofen, Reinhard
University of Zürich
Institute of Plant Biology
Zollikerstrasse 107
CH-8008 Zürich
SWITZERLAND

Ball, Harold L.
Orenco System, Inc.
2826 Colonial Road
Roseburg, OR 97470 USA

Ballew, M. Bruce
Professional Analysis, Inc.
Weldon Spring Site
7295 Highway 94 South
St. Charles, MO 63304 USA

Bates, Edward R.
U.S. Environ. Protection Agency
Natl. Risk Mgmt. Research Lab
26 W. Martin Luther King Drive
Cincinnati, OH 45268 USA

Benemann, John R.
343 Caravelle Drive
Walnut Creek, CA 94598 USA

Berti, William R.
DuPont Co.
Glasgow Site B-301
P.O. Box 6101
Newark, DE 19714-6101 USA

Birch, Linda
University of Zürich
Institute of Plant Biology
Zollikerstrasse 107
CH-8008 Zürich
SWITZERLAND

Bolton, Jr., Harvey
Battelle Pacific Northwest
P.O. Box 999
Richland, WA 99352 USA

Buchanan, Bob B.
University of California at Berkeley
Department of Plant Biology
111 Koshland Hall
Berkeley, CA 94720 USA

Buchs, Urs
University of Zürich
Institute of Plant Biology
Zollikerstrasse 107
CH-8008 Zürich
SWITZERLAND

Carlson, Don
University of California at Berkeley
Department of Plant Biology
111 Koshland Hall
Berkeley, CA 94704 USA

Cevaal, John
Camp Dresser & McKee
1331 17th Street, Suite 1200
Denver, CO 80202 USA

Chasteen, Thomas G.
Sam Houston State University
Dept. of Chemistry
1900 Sam Houston Avenue
Farringtom Building 104
Huntsville, TX 77340 USA

Cornish, James E. (Jay)
MSE, Inc.
P.O. Box 4078
Butte, MT 59702 USA

Cunningham, Scott D.
DuPont Co.
Central R&D
Glasgow Site B-301
P.O. Box 6101
Newark, DE 19714-6101 USA

Farmer, Garry H.
PRC Environmental Mgmt., Inc.
1099 18th St., Suite 1960
Denver, CO 80202 USA

Ferloni, Peter
University of Zürich
Institute of Plant Biology
Zollikerstrasse 107
CH-8008 Zürich
SWITZERLAND

Flynn, Isabelle
University of Zürich
Institute of Plant Biology
Zollikerstrasse 107
CH-8008 Zürich
SWITZERLAND

Garbisu, Carlos
University of California
111 Koshland Hall
Department of Plant Biology
Berkeley, CA 94720 USA

Goldberg, William C.
MSE, Inc.
P.O. Box 4078
Butte, MT 59702 USA

Gorby, Yuri A.
Battelle Pacific Northwest
P.O. Box 999
Richland, WA 99352 USA

Graydon, James W.
Metcalf & Eddy, Inc.
25 Main Street
Chico, CA 95928 USA

Guerin, Turlough F.
Minenco Bioremediation Services
1 Research Avenue
Bundoora, Victoria 3083
AUSTRALIA

Gusek, James
Knight Piesold & Co.
1600 Stout Street, Suite 800
Denver, CO 80202 USA

Hayes, Desmond W.J.
Engineering Research Institute
2320 East San Ramon Avenue
Fresno, CA 93740 USA

Houthoofd, Janet M.
U.S. Environ. Protection Agency
Natl. Risk Mgmt. Research Lab
26 W. Martin Luther King Drive
Cincinnati, OH 45268 USA

Huang, Jianwei W.
DuPont Co.
Glasgow Site B-301
P.O. Box 6101
Newark, DE 19714-6101 USA

Ishii, Takahisa
University of California at Berkeley
Plant Biology
111 Koshland Hall
Berkeley, CA 94720 USA

Jud, Gaudenz
University of Zürich
Institute of Plant Biology
Zollikerstrasse 107
CH-8008 Zürich
SWITZERLAND

Kelly, Robert J.
Dames & Moore
North Sydney
Sydney, New South Wales 2060
AUSTRALIA

Kipps, Jo Anne L.
California Dept. of Water Resources
3374 East Shields Avenue
Fresno, CA 93726 USA

Kovac, Kurt C.
California Dept. of Water Resources
3374 East Shields Avenue
Fresno, CA 93726 USA

Leighton, Terrance
University of California at Berkeley
Molecular and Cell Biology
Berkeley, CA 94720 USA

Levine, Rashalee S.
U.S. Department of Energy
Headquarters (EM-53)
Office of Technology Development
Trevion II Building
12800 Middlebrook Road
Germantown, MD 20874 USA

Lu, Yongming
University of New Mexico
Dept. of Chemical Engineering
209 Farris Engineering Center
Albuquerque, NM 87131-1341 USA

McCarty, Steven L.
Sam Houston State University
Department of Chemistry
1900 Sam Houston Avenue
Farringtom Building, Room 104
Huntsville, TX 77340 USA

Mohagheghi, Ali
National Renewable Energy Lab
Biotechnology Research Branch
1617 Cole Blvd.
Golden, CO 80401-3393 USA

Nagle, Nick J.
National Renewable Energy Lab
Applied Biological Sciences Branch
1617 Cole Blvd.
Golden, CO 80401-3393 USA

Oakley, Stewart M.
California State University
Chico (Dept. of Civil Engineering)
Chico, CA 95929-0930 USA

Owens, Lawrence P.
California State University Fresno
Engineering Research Institute
2320 East San Ramon Avenue
Fresno, CA 93740-0094 USA

Paterek, James R.
Institute of Gas Technology
1700 South Mt. Prospect Road
Des Plaines, IL 60018-1804 USA

Philippidis, George
National Renewable Energy Lab
1617 Cole Blvd.
Golden, CO 80401-3393 USA

Radehaus, Petra M.
Colorado School of Mines
Dept. of Chem. & Geochemistry
Golden, CO 80401-1887 USA

Ramesh, Geetha
Harbor Branch Oceanographic Inst.
5600 U.S. 1 N
Fort Pierce, FL 34946 USA

Reed, Brian H.
California State University, Chico
Department of Civil Engineering
Chico, CA 95929-0930 USA

Rice, Laura E.
Institute of Gas Technology
1700 South Mt. Prospect Road
Des Plaines, IL 60018 USA

Rivard, Christopher J.
National Renewable Energy Lab
Biotech Research Branch
1617 Cole Blvd.
Golden, CO 80401 USA

Scheuering, Joseph
Noranda Minerals Corp.
2501 Catlin, Suite 201
Missoula, MT 59801 USA

Schmidt, Glen C.
MK-Ferguson Co.
Weldon Spring Site
7295 Highway 94 South
St. Charles, MO 63304 USA

Smith, Lawrence A.
Battelle Columbus
505 King Avenue
Columbus, OH 43201-2693 USA

Smith, Nancy R.
California State University
Department of Biological Sciences
Hayward, CA 94542 USA

Srivastava, Vipul J.
Institute of Gas Technology
Biotechnology Research
1700 South Mt. Prospect Road
Des Plaines, IL 60018 USA

Stalder, Verena
University of Zürich
Institute of Plant Biology
Zollikerstrasse 107
CH-8008 Zürich
SWITZERLAND

Tahedl, Harald
University of Zürich
Institute of Plant Biology
Zollikerstrasse 107
CH-8008 Zürich
SWITZERLAND

Updegraff, David M.
Colorado School of Mines
Dept. of Chemistry & Geochemistry
Golden, CO 80401 USA

Varadarajan, Ramesh
Microbial Products, Inc.
Vero Beach, FL 32962 USA

Wang, Tsen C.
Harbor Branch Oceanographic Inst.
5600 U.S. 1 North
Ft. Pierce, FL 34946 USA

Weissman, Joseph C.
Microbial Products, Inc.
Vero Beach, FL 32962 USA

Whiting, Kent
Camp Dresser & McKee
1331 17th Street, Suite 1200
Denver, CO 80202 USA

Wildeman, Thomas
Colorado School of Mines
Dept. of Chemistry & Geochemistry
Golden, CO 80401 USA

Wilkins, Ebtisam
University of New Mexico
Dept. of Chem. & Nuclear Engrg.
209 Farris Engineering Center
Albuquerque, NM 87131-1341 USA

Yee, Andrew
Lawrence Berkeley Laboratory
Division of Earth Sciences
Lawrence Berkeley Laboratory
Berkeley, CA 94720 USA

Yee, Boihon C.
University of California at Berkeley
Department of Plant Biology
111 Koshland Hall
Berkeley, CA 94720 USA

INDEX